Topics in

Enzyme and Fermentation Biotechnology

1

Topics in

Enzyme and Fermentation Biotechnology

1

Editor
ALAN WISEMAN, PH.D., F.R.I.C., M.I.BIOL.
Department of Biochemistry, University of Surrey, Guildford

ELLIS HORWOOD LIMITED
Publisher Chichester

Halsted Press: a division of
JOHN WILEY & SONS INC.
New York · London · Sydney · Toronto

The publisher's colophon is reproduced from James Gillison's drawing of the ancient Market Cross, Chichester.

First published in 1977 by
ELLIS HORWOOD LIMITED, Publisher
Coll House, Westergate, Chichester, Sussex, England

Distributors:

Australia, New Zealand, South-east Asia:

JOHN WILEY & SONS AUSTRALASIA PTY LIMITED
1–7 Waterloo Road, North Ryde, N.S.W., Australia

Canada:

JOHN WILEY & SONS CANADA LIMITED
22 Worcester Road, Rexdale, Ontario, Canada

Europe, Africa:

JOHN WILEY & SONS LIMITED
Baffins Lane, Chichester, Sussex, England

U.S.A., South America and the rest of the World:

HALSTED PRESS a division of
JOHN WILEY & SONS INC.
605 Third Avenue, New York, N.Y. 10016, U.S.A.

Library of Congress Cataloging in Publication Data

Main entry under title:

Topics in enzyme and fermentation biotechnology.

1. Biochemical engineering—Addresses, essays, lectures. I. Wiseman, Alan. [DNLM: 1. Enzymes. 2. Fermentation. QU135 T674]
TP248.3.T66 1977 660'.63 76-25441
ISBN 0-470-98896-7 (Halstead Press)
ISBN 0 85312 051 X (Ellis Horwood)

Filmset and printed Offset Litho in Great Britain by
Cox & Wyman Ltd, London, Fakenham and Reading

Table of Contents

Chapter 1

Introduction to Topics in Enzyme and Fermentation Biotechnology

Dr ALAN WISEMAN, Department of Biochemistry, University of Surrey, Guildford, Surrey

1.1 GENERAL INTRODUCTION TO THE SERIES

The basic precursor to this new series is *Handbook of Enzyme Biotechnology*, ed. Alan Wiseman (Ellis Horwood), 1975. Part I covers the principles of enzyme production and utilization; Part II is a collection of data for industrial application of enzymes. Many specialist topics were mentioned in this source book, but few were subjected to detailed analysis. This first volume of a new series entitled *Topics in Enzyme and Fermentation Biotechnology* reviews a few of these more specialist topics, in detail. Particular attention has been given to the balance of presentation between the aspects of fermentation biotechnology and enzyme biotechnology, and from this relationship the interdisciplinary nature of the study of enzymes and their production by fermentation procedures is apparent. This interdisciplinary approach is a feature also of the series of short post-experiences courses running each July at the University of Surrey, entitled 'Enzyme and Fermentation Biotechnology'. Many of the contributors to Volume I of this new series have lectured on these summer courses.

1.2 INTRODUCTION TO TOPICS IN VOLUME I

The regulation of enzyme biosynthesis, even without recourse to genetic manipulations, remains an important area of interest in relation to the controlled production of microbial enzymes. Induction of the required enzyme is possible in many cases (for review see Wiseman, 1975), especially for enzymes needed for the breakdown of a complex substrate available for growth. Alternatively, in many synthetic pathways subject to feedback repression, the lowering of the concentration of endogenous repressors may be achieved. This may be possible by the exclusion or limitation of repressors, such as certain amino acids, to achieve the increased synthesis of enzymes in the pathway of biosynthesis of this amino acid. Elimination of amino acids from the medium may also de-repress the synthesis of proteases. Similarly, ammonium ions present in the medium will repress the synthesis of urease, and phosphate will repress the synthesis of phosphatase. Limitation of the growth of micro-organisms by restricting the supply of a growth factor may also be useful in lifting repression, as might be the addition of competitive inhibitors of substances causing end-product repression of enzymes of the corresponding pathway.

Catabolite repression, including glucose repression, may also be a problem with some enzymes, especially those involved in the breakdown of complex substrates. The carbon source used may need to be given slowly if it is rapidly utilized, and would normally cause a build up of low molecular weight catabolite repressors. Unlike feedback repression, the action of catabolite (glucose) repressors is through the removal of the cyclic AMP needed for the action of some promoters in the genome. Addition of cyclic AMP may be successful with some organisms, although this would be too expensive an operation for most purposes. It is of great interest that cyclic GMP will reverse the effects of cyclic AMP. Some promoters may have reversed requirements for these cyclic nucleotides (see Chapter 8, Section 6). Studies in continuous culture are of especial interest in this field (see Chapter 2) as is further understanding and control of the fermentation procedure, especially for viscous media (see Chapter 4). New techniques of isolating enzymes on a large scale are of great potential importance (see Chapter 3).

Much of the industrial secrecy surrounding the recent advances in enzyme biotechnology is concerned with the difficulty of adequate patent protection (see Chapter 6). Information on the current use of immobilized penicillin acylase (see Chapter 5) is certainly difficult to document openly. Patenting law, with all its complications, is a difficult subject in relation to micro-organisms and microbial products, especially in view of the changes possibly with the entry of the U.K. to the E.E.C. (Pennant and Wild, 1976). Nevertheless, some new immobilized enzyme applications are well documented now. Glucose isomerase (see Chapter 7) is still the subject of considerable secrecy and commercial rivalry, and developments in the U.K. are awaited with considerable interest. A historical and commercial review on this enzyme by Barker (1975) is most instructive in this connection. A crude enzyme immobilized *in situ* to the bacterial cells seems to be preferred, presumably mainly on an economical basis. These preparations are of relatively high stability, and for example, Novo Industri (the improved Sweetzyme 5) retains 25% of its original activity after 1,050 hours under defined working conditions (probably 65 °C at pH 8·5).

Many new ideas are emerging on the uses of immobilized enzymes and some will require new technologies such as work in non-aqueous media, e.g. in the production of contraceptive steroids. One new idea of considerable interest is the use of enzymes linked to synthetic amphipathic polymers, resulting in an enzymatically active surfactant which is capable of forming stable emulsions with water-immiscible liquids (see patent, West German Offenlegungschrift, No. 2535951).

Medical applications of many enzymes have been suggested, often by use in extra-corporeal shunts or in packages such as liposomes (Gregoriadis, 1973) to prevent immunological responses. The editor's own interest in this connection is discussed in Chapter 8.

REFERENCES

Barker, S. A. (1975), *Process Biochemistry*, **12**(10), 39–40.

Gregoriadis (1973), *New Scientist*, 27th December, 890–893.

Pennant, P. and Wild, G. L. E. (1976), *Chemistry and Industry*, 17th January, 53–56.

Wiseman, A. (1975), in *Enzyme Induction*, 1–26. Edit. by Parke, D. V., Plenum Press, London and New York.

Chapter **2**

Regulation of Enzyme Synthesis in Continuous Culture

Dr J. MELLING, Microbiological Research Establishment, Porton Down, Salisbury, Wiltshire

2.1 INTRODUCTION

2.1.1 The Technique of Continuous Culture

Growth of micro-organisms in continuous culture takes place under steady-state conditions and the environmental factors, which can vary markedly during the 'growth cycle' in batch culture, are maintained constant. Furthermore, such culture parameters as pH, temperature, nutrient concentration and growth rate can be altered readily for experimental purposes.

The theoretical basis of continuous culture was established by the papers of Monod (1950) and Novick and Szilard (1950). Later, Herbert, Elsworth and Telling (1956) expanded the fundamental principles and showed that the theoretical predictions could be validated experimentally. It is not considered necessary to reiterate in detail the, by now well established, theory of continuous culture; and reference to the above mentioned authors and to the papers of Herbert (1958) and Powell (1956, 1958, 1972) should provide an adequate theoretical treatment. Let it suffice to say that the specific growth rate of an organism (μ) is determined by the concentration (S) of some growth-limiting substrate thus:

$$\mu = \mu_m \frac{S}{K_s + S}$$

Where μ_m is the maximum growth rate in the presence of excess substrate and K_s a saturation constant. Thus μ can have different values up to μ_m depending upon the supply of growth-limiting substrate which can be controlled by the rate of flow of fresh medium into the growth vessel.

Two types of continuous culture have been used: (1) the chemostat in which it is the rate of supply of fresh medium which controls the growth rate through some feedback system, which is used to control the flow rate (Novick, (2) the turbidostat is a device in which it is the organism concentration, through some feedback system, which is used to control the flow rate (Novick, 1955; Munson 1970). When μ is well below μ_m there are only small variations in organism and substrate concentrations for minor changes in dilution rate and it is here that chemostat operation is successful. However, as μ approaches

μ_m a small change in dilution rate (see Herbert, 1958, Figure 2.3) can effect a large variation in organism and substrate concentrations and thus the direct control of the organism concentration allowed by the turbidostat gives a more stable system.

Because it is possible to obtain stable steady-state populations of micro-organisms under a variety of conditions, continuous culture has proved to be a valuable tool in the study of microbial physiology. However, it is only during the last five years or so that an increase in the application of the technique to enzyme regulation and biosynthesis has been seen.

2.1.2 The Bacterial Environment

In order to flourish in natural ecosystems, bacteria have to cope with wide fluctuations in such factors as pH, temperature and nutrient availability. Accordingly, they have acquired a high degree of plasticity in their physiological responses to changes in the environment. Since the expression of the genotype into the phenotype is largely mediated by enzymes it is scarcely surprising that changes in the growth environment can result in wide variations in enzyme activity.

It is by now well established that the composition and morphology, as well as particular characteristics of micro-organisms, can change markedly according to the growth conditions (Herbert, 1961; Ellwood and Tempest, 1972; Melling and Brown, 1975) and these changes reflect the regulation of enzyme synthesis and activity. Such regulatory mechanisms may operate at the cytoplasmic level by controlling enzyme activity through a system of end-product inhibition. Here the end-product of a series of reactions may affect the activity of an enzyme involved at an earlier point in the pathway as a result of the allosteric nature of some enzyme molecules (Monod, Changeux and Jacob, 1963). This type of control, although rapid, has the disadvantage for the organism that an unnecessarily large amount of enzyme may be produced only to have its activity reduced later. Consequently it is useful for organisms to have the capacity to regulate enzyme synthesis at the genetic level in order to control enzyme production according to the conditions pertaining. This is achieved by such regulatory mechanisms as induction and repression (Jacob and Monod, 1961) and catabolite repression (Magasanik, 1961). It is with the control of enzyme synthesis, rather than activity, that this chapter is concerned. When using a micro-organism for production of some material it is required to disrupt the regulatory mechanisms and, as it were, drive the organism into a metabolic corner.

The technique of continuous culture is one way of achieving this objective. As well as producing phenotypic changes in microbial physiology, growth in continuous culture can be used to exert selective pressure for the isolation of mutants. In addition it allows growth conditions in natural environments to be simulated, which batch culture does not. Apart from ecological studies in general this may allow a greater understanding of microbial pathogenicity to be attained.

The majority of work using continuous culture has been done by maintaining steady-state conditions in different situations so that the responses of the organism could be correlated with particular environmental conditions. However, the transition between one steady state and another is an important physiological step. Thus the study of enzyme activities in such transition states should prove of great interest and some work has already been reported (Dean, 1972; Carlsson and Elander, 1973; Ellwood, Hunter and Longyear, 1974). Furthermore, as Dean (1972) mentioned, the attainment of a steady state with respect to cell concentration is no guarantee of a steady state as far as some enzymes are concerned.

2.1.3 Bibliography

For readers interested in aspects of continuous culture outside the scope of this chapter the following series of published books may prove useful.

International Continuous Culture Symposia:

1st *Continuous Cultivation of Microorganisms*, Editor: I. Malek, Czech. Acad. Sci. Prague, 1958.

2nd *Continuous Cultivation of Microorganisms*, Editors: I. Malek, K. Beran and J. Hospodka, Czech. Acad. Sci. Prague, 1964.

3rd *Microbial Physiology and Continuous Culture*, Editors: E. O. Powell, C. G. T. Evans, R. E. Strange and D. W. Tempest, H.M.S.O. London, 1967.

4th *Continuous Cultivation of Microorganisms*, Editors: I. Malek, K. Beran, Z. Fencl, V. Munk, J. Ricica and H. Smrckova, Academia, Prague, 1969.

5th *Environmental Control of Cell Synthesis and Function*, Editors: A. C. R. Dean, S. J. Pirt and D. W. Tempest, Academic Press, London, 1972.

6th *Continuous Culture: Applications and New Fields*, Editors: A. C. R. Dean, D. C. Ellwood, C. G. T. Evans and J. Melling, Ellis Horwood, Chichester, 1976.

Continuous Cultivation of Microorganisms—reviews, J. Ricica:

1970 *Folia Microbiol.* **15**, 377–416.
1971 *Folia Microbiol.* **16**, 389–415.
1972 *Folia Microbiol.* **17**, 398–432.
1973 *Folia Microbiol.* **18**, 418–448.
1974 *Folia Microbiol.* **19**, 397–436.

2.2 GROWTH MEDIUM COMPOSITION

2.2.1 Growth-limiting Nutrients

In a large number of continuous culture studies emphasis has been placed upon the growth medium composition with respect to a particular growth-limiting nutrient, since variations in the supply of such a nutrient affect many characteristics of the culture. However, the levels of other non-limiting nutrients may also result in significant alterations to an organism's metabolism. In this section such effects will be considered together with those of specific growth-limiting nutrients.

In order to accommodate their metabolism to the fluctuations in nutrient availability which occur in natural environments, organisms can regulate the synthesis of particular enzymes involved in the assimilation of nutrients. Alternatively, they may increase the availability of growth-limiting nutrients by secreting enzymes such as proteases, amylases and phosphatases to solubilize otherwise inaccessible substrates.

The assimilation of ammonia in micro-organisms may proceed either directly by the formation of glutamate from 2-oxoglutarate using glutamate dehydrogenase (I) or by way of the glutamine synthetase (II)/glutamate synthase route (III) (Tempest, Meers and Brown, 1970) as shown in Figure 2.1. The data in Table 2.1 clearly indicate how the levels of these enzymes vary

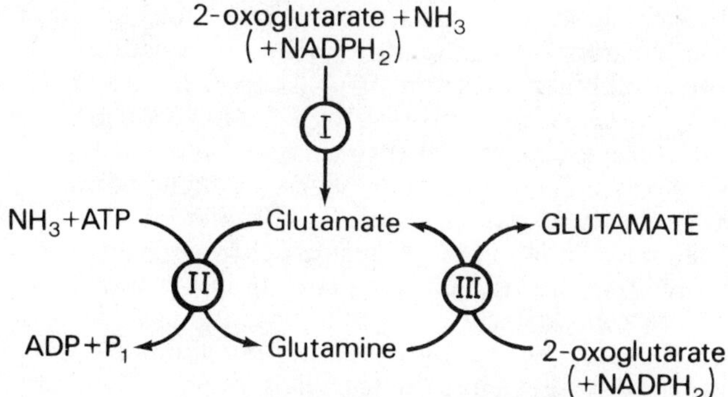

FIG. 2.1 Pathways of ammonia assimilation in *K. aerogenes.*

Overall reactions:

(I) Glutamate dehydrogenase:

$$NH_3 + \text{2-oxoglutarate} + NADPH_2 \rightarrow \text{Glutamate} + NADP + H_2O.$$

(II and III) Glutamine synthetase plus glutamate synthase:

$$NH_3 + \text{2-oxoglutarate} + NADPH_2 + ATP \rightarrow \text{Glutamate} + ADP + P_i + NADP.$$

[*Reproduced with permission from D. W. Tempest and O. M. Neijssel. In* 'Continuous Culture: Applications and New Fields'. *A. C. R. Dean, D. C. Ellwood, C. G. T. Evans and J. Melling (eds), Ellis Horwood, Chichester, 1976.*

Table 2.1 The Influence of the Limiting Substrate on the Levels of Enzymes Responsible for Glutamate Synthesis using Ammonia as Sole Nitrogen Source. [*Data from Brown, Macdonald-Brown & Stanley (1973); Brown (1976).*]

Organism	*Limiting nutrient*	*Enzyme activities* (nmole/min/mg protein) *Glutamine synthetase*	*Glutamate dehydrogenase*	*Glutamate synthetase*
Klebsiella aerogenes	Carbon	<1	671	<1
	Ammonia	91	<1	39
Pseudomonas aeruginosa	Carbon	<1	48*	15
	Ammonia	171	3*	30

* Refers to the total of NAD- and NADP-linked enzymes.

according to the ammonia content in the culture. It is scarcely surprising that the assimilatory route via glutamine synthetase and glutamate synthase represents a high-affinity uptake system with a K_m for ammonia well below that of the glutamate dehydrogenase system. However, the high affinity uptake system requires 1 mole of ATP for each mole of ammonia taken up and thus in the presence of excess ammonia the glutamate dehydrogenase route is less expensive in terms of energy. Such control of enzyme synthesis depending upon the availability of ammonia, and other nitrogen sources has been described in a number of organisms (Brown, 1976). It is, however, interesting that not all organisms possess the high-affinity uptake system just described and a number of these have been found in which the levels of glutamate dehydrogenase were increased many-fold in nitrogen-limited compared with carbon-limited cultures (Brown, Macdonald-Brown and Mears, 1974; Griffith and Carlsson, 1974; Burn, Turner and Brown, 1974; Brown and Johnson, 1970). A possible explanation for the ability of some organisms to utilize glutamate dehydrogenases (Apparent K_m for ammonia 2·0–5·0 mM) to assimilate ammonia under ammonia-limiting conditions (about 2 mM ammonia) was given by Griffith and Carlsson (1974). They suggested that since coupling of glutamate dehydrogenase with an aminotransferase can lower its K_m for ammonia (Fahien, Lun-Yu, Smith and Happy, 1971) this might facilitate uptake of ammonia. In support of this proposal, which would require high levels of both glutamate dehydrogenase and aminotransferase, they cited the increased levels of glutamate dehydrogenase which occurred in their ammonia-limited cultures and the large amounts of glutamate-oxalacetate aminotransferase which were detected in the strain of *Streptococcus sanguis* which they used.

In organisms possessing a high-affinity uptake system for nitrogen it is not only the production of glutamine synthetase which is regulated by ammonia (Prival, Brenchley and Magasanik, 1973) but also that glutamine synthetase activates the transcription of genes for histidase (Tyler, De Leo and Magasanik, 1974) proline oxidation and tryptophan permease (Magasanik, *et al.*, 1974). In addition Kavanagh and Cole (1976) have shown that nitrite reductase levels in *Escherichia coli* were high under conditions of nitrogen limitation, either by ammonia or nitrite, but enzyme levels were repressed when these nutrients were supplied in excess. They concluded that

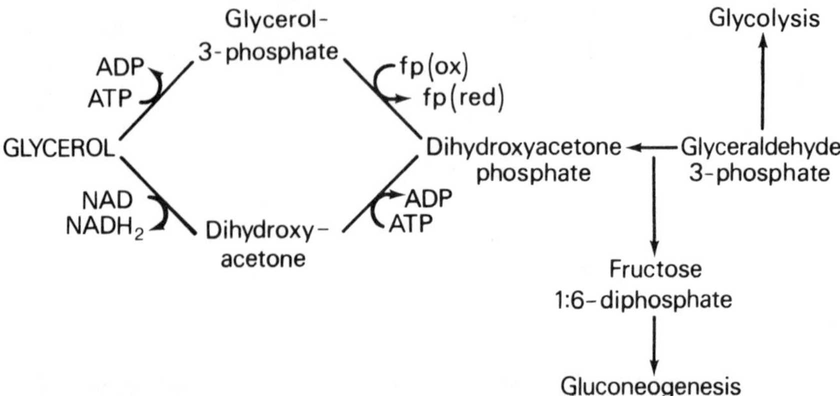

FIG. 2.2 Dual pathways of glycerol assimilation in *Klebsiella aerogenes*. [*Reproduced with permission from O. M. Neijssel, S. Hueting, K. J. Crabbendam and D. W. Tempest,* Arch. Microbiol., *1975*, ***104****, 83–87.*

the regulation of nitrite reductase was solely by ammonia repression rather than by substrate induction.

As well as enzymes involved in nitrogen uptake, considerable variation has also been found in those responsible for assimilation of the carbon and energy sources. Neijssel, Hueting, Crabbendam and Tempest (1975) have described the regulation of enzymes involved in uptake of glycerol in *Klebsiella aerogenes* (Figure 2.2). Two pathways for assimilation of glycerol are available. The first involves conversion of glycerol to glycerol-3-phosphate by glycerol kinase and then to dihydroxyacetone phosphate by glycerol-phosphate dehydrogenase. The second also results in the formation of dihydroxyacetone phosphate, but involves glycerol dehydrogenase and then dihydroxyacetone kinase. The glycerol dehydrogenase pathway is induced when glycerol is present in excess of growth requirements and the first enzyme in this pathway has a low affinity for glycerol (apparent $K_m = 1\text{–}5 \times 10^{-2}$ M). The alternative, glycerol kinase, has a high affinity for glycerol (apparent $K_m = 1\text{–}5 \times 10^{-6}$ M). Glycerol kinase was found to be synthesized in glycerol-limited environments.

The control of glucose uptake and metabolism in *Pseudomonas aeruginosa* which was described by Whiting, Midgley and Dawes (1976) provides further evidence of the variations in enzyme levels in response to particular nutrient limitations as well as clearly defining the elegant control systems for glucose metabolism in this organism. Figure 2.3 indicates the metabolic pathways both intracellularly and extracellularly for glucose metabolism and Table 2.2 shows the variations in glucose catabolizing enzymes with changes in growth limiting nutrient. The authors concluded that both glucose dehydrogenase and gluconate dehydrogenase act extracellulary to produce gluconate and 2-oxogluconate and that these enzymes are regulated independently and repressed by growth on organic acids such as citrate. Glucose, gluconate and 2-oxogluconate are transported by separate transport systems which are induced by their particular substrates. These transport systems are also

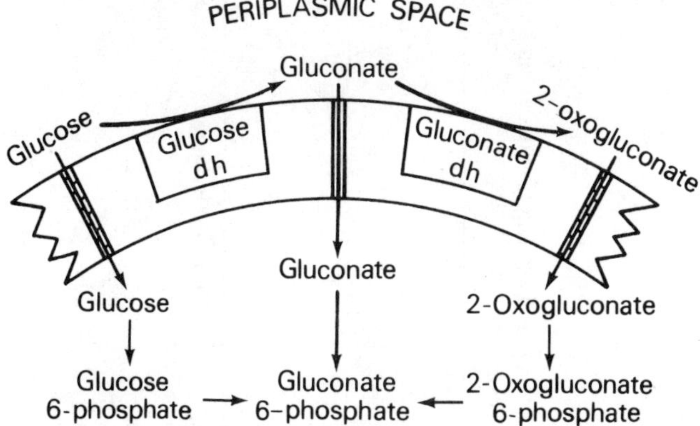

FIG. 2.3 Extracellular and intracellular metabolism of glucose by *P. aeruginosa* (dh = dehydrogenase). [*Reproduced with permission from P. H. Whiting, M. Midgley and E. A. Dawes,* J. gen. Microbiol., *1976*, **92**, *304–310.*

Table 2.2 Effect of Limiting Nutrient on the Glucose-catabolizing Enzymes of Glucose-grown *P. aeruginosa* PAO1. [*Reproduced with permission from P. H. Whiting, M. Midgley and E. A. Dawes (1976), in* J. gen. Microbiol., **92**, *304–310*.

	Specific activity (μmol substrate converted/h/g protein)						
Limiting nutrient	*Glucose* dh	*Gluconate* dh	*Gluconate kinase*	*2-Oxogluconate metabolizing enzymes*	*Hexokinase*	*Glucose 6-phosphate* dh	*Entner-Doudoroff enzymes*
Nitrogen (NH_4^+)	10·8 13·6	69·8 59·6	2·4 2·1	2·7 2·9	22·7 19·2	29·9 34·6	19·7 14·6
Carbon (glucose)	8·5 9·1	11·1 15·0	0·7 0·7	0·3 0·3	31·0 28·3	68·0 57·2	15·1 16·7

Dilution rate, 0·41 h^{-1}. The two values quoted are the mean of duplicate determinations performed on two separate culture samples from each steady state.
dh = dehydrogenase.

repressed by growth on organic acids and the glucose transport system is repressed by gluconate.

Under nitrogen-limited conditions with glucose as sole carbon source (Table 2.2) a considerable proportion of the glucose is metabolized *via* the extracellular pathway and gluconate and 2-oxogluconate accumulate in the medium since they are produced in excess of the capacity of the transport systems. However, under glucose-limitation the extracellular enzymes are repressed together with the associated transport systems. In addition the levels of the glucose transport system, hexokinase and glucose-6-phosphate dehydrogenase are increased. Thus the available glucose is taken up as quickly as possible. A similar scheme has been proposed for *Pseudomonas fluorescens* (Eisenberg, Butters, Quay and Friedman, 1974) and the levels of the particular enzymes should also respond to modification by chemostat culture.

The levels of a number of other microbial enzymes have been found to respond to changes in growth-limiting nutrients. Although these responses have not yet been explained so completely in terms of microbial physiology they are nevertheless of interest.

The penicillin amidase content of *Bacillus megaterium* was found to increase 3–5 fold when the organism was grown under sulphur-limited rather than nitrogen-limited conditions (Acevedo and Cooney, 1973) and the authors suggested that this might result from a depletion of sulphur-containing proteins. Similarly Robinson and Tempest (1973) observed a decrease in the sulphur content of *K. aerogenes* and of their cell-wall and soluble proteins in sulphate-limited chemostat culture.

Perhaps more specifically, repression of a number of enzymes both intra- and extracellular has been observed to depend upon the levels of nitrogen and carbon sources in the growth media: with an increase in enzyme synthesis when the repressor was made the growth-limiting nutrient. The nitrogenase of *Klebsiella pneumoniae* showed repression by ammonia when grown under sulphate limitation and catabolite repression by the carbon source when grown under nitrogen limitation (Tubb and Postgate, 1973). A similar effect of ammonium ions in repressing nitrogenase was reported by Hovius, Conway and Ganze (1973); also in sulphate-limited culture. Repression by both nitrogen and carbon sources has been observed for other enzymes. Heineken and O'Connor (1972) investigated three catabolite repressible enzymes produced by *Bacillus subtilis*: an α-amylase and a neutral and an alkaline protease. α-Amylase was repressed whenever glucose was present in excess whereas both proteases were repressed completely only by glucose and amino acids. At low amino acid concentrations glucose had little or no repressing activity on the proteases and maximum activity was obtained using a glucose/casamino acids medium under nitrogen limitation at a dilution rate of 0·3 hr^{-1}. Production of an α-amylase was also investigated by Meers (1973) using *Bacillus licheniformis.* In carbon-limited cultures no inducers were found to be needed and the author considered that the prime factors involved in modifying enzyme production were catabolite repression and the specific growth rate. Payne (1972) found that the activity of various intracellular dipeptidases of *E. coli* grown at a dilution rate of 0·18 hr^{-1} varied with the growth limiting nutrient. Potassium limitation resulted in high enzyme levels, but these were low under carbon limitation. Tripeptidase activity was also highest in potassium-limited cultures, but lowest with either nitrogen- or magnesium-limitation. The author suggested that the levels of these enzymes might relate to the degree of protein turnover in the organisms.

Carbon limitation was also found to de-repress the synthesis of pullulanase by *K. aerogenes* (Dean, Gray, Hope and Johnson, 1972) using maltose, maltotriose or pullulan as carbon sources. Little or no enzyme was produced under other limitations and although the authors could envisage the requirement for this enzyme when pullulan or amylopectin were used as carbon sources they could not explain its function in organisms growing on maltose. Using the same organism, Bolton and Dean (1972) studied the production of phosphatases as a function of dilution rate and growth limiting nutrient. Their results shown in Table 2.3 indicate the wide variations in enzyme

Table 2.3 Effect of Growth-limiting Nutrient and Dilution Rate (D) on Acid (pH 4·6) and Neutral (pH 7·1) Phosphatase Specific Activities of Intact Cells of *K. aerogenes*. [*Reproduced with permission from P. G. Bolton and A. C. R. Dean (1972).* Biochem. J. **127**, *87–96*.] Glucose was used as the sole carbon source throughout and the cells were grown in chemostat culture in the given nutrient-limited conditions until steady-state conditions were achieved.

D (h^{-1})	*Assay* pH	*Limiting nutrient*	*Phosphatase specific activity* C	NH_4^+	K^+	Mg^{2+}	SO_4^{2-}
0·4	4·6		100	88	75	60	38
	7·1		50	14	46	52	58
0·7	4·6		76	69	54	43	30
	7·1		30	19	38	39	55
1·0	4·6		35	47	36	31	26
	7·1		11	15	21	22	35

yields associated with changing growth conditions. They also observed that compared with glucose as carbon-limiting growth material other sugars not only altered the activity but also changed the pH profile of these enzymes.

The use of phosphate limitation for production of an alkaline phosphatase from *E. coli* is well established and was discussed by Torriani (1960). More recently Sayer (1968) showed that maximum levels of this intracellular enzyme were obtained at low dilution rates under phosphate limitation when the phosphate concentration in the culture was low.

A number of workers have investigated β-galactosidase synthesis in continuous culture and although much of this has been concerned with the use of the chemostat for mutant selection (see Section 2.3.1) some aspects concerning phenotypic variation are of interest here. Smith and Dean (1972) observed that the variations of enzyme yield with dilution rate in *K. aerogenes* were dependent upon the growth limiting nutrient. With lactose as the limiting nutrient the enzyme level passed through a sharp maximum at $D = 0{\cdot}2$–$0{\cdot}25$ hr^{-1} whereas in K^+-, NH_4^+-, Mg^{2+}-, SO_4^{2-}- and PO_4^{3-}-limited systems the specific activity increased with increasing dilution rate. Furthermore, the variation in enzyme activity with nutrient limitations other than lactose was small compared with that due to changing from carbon limitation to limitation by other nutrients. The major factor determining the β-galactosidase activity (a 20-fold increase with lactose limitation compared with others) was the relative concentration of lactose in the medium.

Shortly after changing from batch to continuous culture at $D = 0{\cdot}5\ \text{hr}^{-1}$ the β-galactosidase activity reached a quasi steady state which persisted for about 15 generations after which it rose steadily until it attained a maximum value (hyperactivity) after about 100 generations. During the quasi steady state the enzyme could be further induced by adding the non-metabolizable inducer methyl-β-D-thiogalactoside (MTG) or repressed if glucose instead of lactose was the carbon source. In contrast, the specific activity of bacteria exhibiting hyperactivity was only slightly reduced by glucose and addition of MTG failed to induce more enzyme. The repression of this enzyme (and others) by glucose and various carbon sources is partially explicable in the light of the data presented by Perlman, de Crombrugghe and Pastan (1969) on the role of cyclic AMP in synthesis of β-galactosidase in *E. coli.* Some aspects of cyclic AMP involvement in bacterial metabolism have been reviewed by Robinson, Butcher and Sutherland (1971). In chemostat culture of *E. coli* constitutive for β-galactosidase, Mandelstam (1962) reported that under nitrogen limitation the enzyme was repressed by glucose, glycerol, gluconate and lactate but switching to carbon limitation using these compounds gave rise to increased β-galactosidase activity.

The effect of different carbon sources on the production of l-asparaginase by *Erwinia carotovora* was examined by Callow, Capel and Evans (1970, 1971) who found that using citric acid or glycerol as carbon sources resulted in bacteria having specific activities of 26 and 30 units/mg cell dry weight respectively whilst mannitol-grown cells had a specific activity of 3·4: chemostat cultures were grown at a $D = 0{\cdot}2\ \text{hr}^{-1}$.

Demain (1971) in reviewing some environmental effects on the production of various metabolites emphasized the requirement for preventing inhibition or repression of early biosynthetic enzymes by feeding growth-limiting concentrations of some final product of the particular biosynthetic pathway. While enzyme inhibition is not the concern of this article, two examples of the removal of enzyme repression are of interest. An early report by Gorin and Maas (1957) showed that the production of the enzyme ornithine transcarbamylase which catalyses the conversion of ornithine to citrulline was repressed by arginine. Thus by using arginine as the growth limiting nutrient in a continuous culture high levels of the enzyme were obtainable. Feeding limiting amounts of threonine to a mutant lacking in homoserine dehydrogenase prevents concerted feedback inhibition of aspartokinase and allows lysine to accumulate.

Repression of A-type cytochrome in *Saccharomyces cerevisiae* was observed when glucose was present in excess in chemostat culture, but synthesis of this material resumed when glucose was growth limiting (Akbar, Rickard and Moss, 1974). A more detailed study of the organization of respiratory chains of *E. coli* grown either sulphate- or glycerol-limited was carried out by Poole and Haddock (1975). Glycerol-limited cells contained cytochromes b_{556}, b_{562} and O as well as ubiquinone and low concentrations of menaquinone. Sulphate limitation caused the additional synthesis of cytochromes d, a, b_{558} and C_{550}, while the level of ubiquinone was reduced and that of menaquinone was barely detectable. Sulphate-limited cells also lacked site I phosphorylation. Although these authors could not entirely exclude the possibility that the effects of sulphate limitation were due to an excess of the carbon source it seems unlikely. Light (1972) reported that the changes in electron-transport chain and energy conservation resulting from iron- or sulphur-limitation in *Candida utilis* were not mimicked by growth limitation using nitrogen, magnesium or phosphorus, thus refuting an earlier suggestion that such changes resulted from some type of catabolite repression (Katz, Kilpatrick and Chance, 1971).

2.2.2 Toxic Substrates

The technique of continuous culture is well suited to the study of enzyme induction by and utilization of materials toxic to micro-organisms. In batch culture experiments there is always the difficulty of adding too small an initial concentration of inducer, which may be degraded by the basal enzyme level, so that in fact the culture grows in the presence of little or no inducer. Alternatively, too high a concentration of such a substrate can render the culture sterile. In continuous cultures, however, steady-state concentrations of inducer can be maintained at any desired level.

Mathematical models of inhibitor effects have been described by Van Uden (1967), Andrews (1968), Aiba, Shoda and Nagatani (1968) and Edwards (1970) concerning the relationships between growth and inhibition.

Phenol has been used as a limiting nutrient by several workers (Rieche, Hilgetag, Martini and Lorenz, 1964; Evans and Kite, 1964; Jones and Carrington, 1972) in studying the detoxification of phenolic waste, but so far no details of the variations in enzyme activities associated with this substrate are available.

There is considerable interest in the growth of organisms on C_1 compounds particularly related to single cell protein production. Harder, Attwood and Quale (1973) investigated the effect on the activity of various enzymes of a *Hyphomicrobium* strain grown in batch culture using either methanol or ethanol as the sole carbon and energy source. They found considerable differences in the levels of enzymes involved in utilization of these two compounds depending upon which compound was used. Continuous culture studies on the utilization of methanol (Levine and Cooney, 1973; Asthana, Humphrey and Moritz, 1971; Snedecor and Cooney, 1974; Takada, Sawada, Sawada, Amano and Terui, 1972) have been carried out, but were concerned with growth yield rather than enzyme regulation *per se*. Mateles (1975) reported growth of *Pseudomonas* C on methanol, formaldehyde and formate under carbon limitation in continuous culture, but this organism was unable to grow on formaldehyde or formate in batch culture. An examination of the microbial physiology of organisms in these situations would clearly be of interest.

Toxic substrates which are not growth-limiting nutrients have also been examined. Herbert and Phipps (1974) reported the induction of catalase in *B. megaterium* growing in a chemostat when hydrogen peroxide was added to the growth medium (mannitol-salts, carbon-limited). It was found that the catalase content of the cells increased with increasing H_2O_2 concentrations in the growth medium. When 0·3 M H_2O_2 was used the catalase represented about 1% of the cell dry weight, an increase of 35- to 60-fold over the basal level. However the concentration of H_2O_2 in the culture was less than 10^{-4} *M*. The authors stated that the effect was not due to selection of high-catalase mutants nor to a high oxygen tension produced by decomposition of the H_2O_2. Hyperbaric oxygen did increase the catalase content, but only 6-fold. They suggested that the slight increase in catalase due to oxygen may have resulted from the formation of superoxide and its subsequent conversion to H_2O_2 by superoxide dismutase. Other enzymes examined (14 in all) were unaffected by feeding H_2O_2, although the ferredoxin levels increased 5-fold. Catalase induction by H_2O_2 was not subject to catabolite repression, at least when mannitol was the carbon source and cultures were nitrogen-limited, as shown in Table 2.4 (Herbert, personal communication).

Table 2.4 Induction of Catalase by Hydrogen Peroxide in *B. megaterium* Growing in Nitrogen-limited Continuous Culture $D = 0{\cdot}44$ hr^{-1}. [*O. Herbert, unpublished results.*]

H_2O_2 *conc. in inflowing medium* (M)	*Catalase content of cells (% of dry weight)*
0·00	0·015
0·56	0·215
0·106	0·391
0·208	0·919
0·310	0·907

Some effects on enzyme activity of adding various toxic agents to both chemostat and turbidostat cultures were studied by Dean and his colleagues. The addition of barbitone to turbidostat cultures of *K. aerogenes* was found by Dean and Moss (1971) to produce damped oscillations in growth rate until a new steady state was attained. The activity of several enzymes also fluctuated and when the new steady state was established the activities of: glucokinase, glucose phosphate isomerase, glucose dehydrogenase, glucose-6-phosphate dehydrogenase and 6-phosphogluconate dehydrogenase were all found to have increased. The authors suggested that addition of barbitone produced initially a generalized inhibition of glucose metabolism and that this was counteracted by an increase in the activities of the various enzymes. In chemostat cultures of *K. aerogenes* it was found (Dean and Rogers, 1969) that adding urea to maltose-limited cultures also led to a compensatory

Table 2.5 Effect of Added $NaNO_2$ on the Levels of Cytochrome Oxidase in *P. aeruginosa* grown Carbon-limited in a Citrate-salts Medium $D = 0{\cdot}42$ hr^{-1}. [*J. Melling and B. W. Phillips, unpublished results.*]

NO_2 *conc.* (mM)	*Enzyme* *u/ml *culture*	*Specific activity* u/g *protein*
0	0·182	167
12	0·259	320
18	0·231	286
29	0·091	225
44	Culture washed out	

* mM O_2 transferred/min using the TMPD/ascorbate assay described by

increase in the α-glucosidase activity of the cells. Variations in enzymes of glucose metabolism also accompanied the addition of 9-aminoacridine, sulphanilamide or 2:4-dinitrophenol to chemostat cultures of *K. aerogenes* (Rogers, 1968).

The addition of nitrite to citrate-limited cultures of *P. aeruginosa* (Table 2.5) produced an increase in the specific activity of cytochrome oxidase (which also functions as a nitrite reductase in this organism) (Melling and Phillips, unpublished results). It is possible that this increase in enzyme activity may be a response to the toxic effects of nitrite.

These various examples make it clear that it is not only limiting nutrients which can affect enzyme activity but that other materials which exert some adverse physiological effect may produce phenotypic variations as the organisms attempt to counter such toxic effects.

2.2.3 Mixed Substrates

It was Monod (1942) who first observed that when batch cultures of bacteria have two carbon and energy sources available simultaneously one may be used until it is exhausted and then, following a lag period, the other is metabolized. This behaviour was termed diauxic growth and Paigen and Williams (1970) stated that two conditions are necessary for it to occur. (1) Adaptation to the secondary carbon source is completely suppressed in the presence of the first one; (2) that adaptation to the second substrate occurs under non-gratuitous conditions: otherwise combinations of substrates may be used simultaneously or sequentially without a lag period.

Continuous culture studies using mixed substrates have been confined to carbon and energy sources. Also, very little information is available concerning the effects on enzyme activity, although inferences can be drawn in some cases where direct measurements are lacking.

Mateles, Chian and Silver (1967) found that *E. coli* and *Ps. fluorescens* were able to use both glucose and fructose simultaneously at low dilution rates, but at high dilution rates much of the fructose remained unused. In contrast *Saccharomyces cerevisiae* exhibited poor utilization of fructose even at low dilution rates. Utilization of glucose/lactose mixtures (Silver and Mateles, 1969) by *E. coli* followed a similar pattern to its growth on glucose/fructose. With prolonged growth mutant selection occurred and this aspect will be discussed later. An exception to the pattern just described was reported by Standing, Fredrickson and Tsuchiya (1972). A strain of *E. coli* which showed diauxic growth on glucose and xylose in batch culture utilized both sugars at low and high dilution rates in continuous culture.

The ability of organisms to utilize two substrates simultaneously in continuous culture presumably indicates the lack of either catabolite repression or catabolite inhibition. This results from the low level of the growth-limiting nutrient particularly at low dilution rates. The discrepancies observed between different organisms and different substrates may reflect the efficiency of catabolite repression in the various cases.

Studies on the effect of mixed substrates on enzyme activities were made by Ng and Dawes (1973) who used ammonia-limited cultures to examine the effects of combinations of glucose and citrate. It had previously been shown that the organic acid was used preferentially (Hamilton and Dawes,

Table 2.6 Effect of Relative Citrate: Glucose Concentrations in Inflowing Medium on Steady-state Enzymic Activities of Citrate-grown *P. aeruginosa*. [*Reproduced with permission from F. M.-W. Ng and E. A. Dawes (1973)*. Biochem. J. **132**, *129–140*.]

The dilution rate was 0·25 h^{-1}. The values recorded are for the steady states established at each substrate concentration ratio.

	Specific activity (μmol/h per mg of N)								
Citrate: glucose conc. ratio in medium (mM)	*Glucose 6-phosphate dehydrogenase*	*Hexokinase*	*6-Phosphogluconate dehydrogenase*	*Gluconokinase*	*Entner-Doudoroff enzymes*	*Glucose dehydrogenase*	*Gluconate dehydrogenase*	*Isocitrate dehydrogenase*	*Aconitase*
75:0	17·5	8·5	0·6	—	2·0	1·2	9·7	302·3	26·0
75:0·5	16·5	13·1	0·4	—	3·8	1·7	16·0	269·4	30·0
75:1	14·7	9·4	0·6	—	4·3	0·9	18·7	295·5	33·2
75:2	19·7	16·0	0·6	—	5·2	1·9	23·6	274·6	31·5
75:4	22·4	15·4	2·7	0·1	5·4	1·6	30·7	290·6	27·2
75:6	30·3	19·5	3·9	0·3	7·6	2·2	41·1	358·2	40·9
75:8	73·6	29·8	18·6	1·0	16·8	2·7	38·9	332·9	35·8
75:10	83·1	35·0	20·1	4·2	17·4	2·4	64·5	325·8	36·0
75:15	94·2	37·7	20·6	3·8	21·8	2·5	74·9	352·6	40·4
75:20	98·7	37·4	24·5	4·1	28·8	2·9	78·1	357·9	43·3
75:30	101·4	39·2	24·7	6·7	30·3	3·1	86·9	338·2	44·0
60:30	116·5	44·3	28·1	4·9	32·7	3·5	78·8	276·2	43·3
45:30	147·6	45·5	30·1	5·2	34·1	11·1	83·1	277·4	43·8

Table 2.7 Effect of Relative Glucose:Citrate Concentrations in Inflowing Medium on the Steady-state Enzymic Activities of Glucose-grown *P. aeruginosa*. [*Reproduced with permission from F. M.-W. Ng and E. A. Dawes (1973)*. Biochem. J. **132**, *129–140*.]

The dilution rate was 0·25 h^{-1}. The values recorded are for the steady states established at each concentration ratio.

	Specific activity (μmol/h per mg of N)								
Glucose: citrate conc. ratio in medium (mM)	*Glucose 6-phosphate dehydrogenase*	*Hexokinase*	*6-Phospho-gluconate dehydrogenase*	*Glucono-kinase*	*Entner-Doudoroff enzymes*	*Glucose dehydro-genase*	*Gluconate dehydro-genase*	*Isocitrate dehydro-genase*	*Aconitase*
45:0	110·1	22·4	19·8	4·7	27·9	17·7	292·6	143·7	26·9
45:4	107·0	21·8	19·5	3·4	22·9	11·2	312·3	147·0	26·3
45:8	119·6	21·7	21·9	4·1	28·1	14·0	301·9	176·7	31·4
45:16	72·9	19·9	13·0	4·4	13·9	14·0	211·2	181·0	26·1
45:32	76·9	16·6	11·2	3·6	8·7	11·2	198·7	176·6	24·5
45:64	40·1	12·5	2·2	3·8	3·0	4·4	46·3	180·7	26·5
0:64	5·8	3·2	1.0	1·9	0·6	4·5	13·9	174·0	26·8

1959) and gave a higher growth rate than glucose in batch culture. When the growth medium for continuous culture contained 75 mM citrate and glucose was subsequently introduced into the culture feed, a glucose concentration of 6–8 mM (residual 4·2 mM) had to be attained before a substantial increase in the specific activities of enzymes of glucose metabolism occurred as shown in Table 2.6. In a converse experiment where citrate was introduced during growth on 45 mM glucose there was a rapid induction of the citrate-transport system and the maximum response occurred with 8 mM citrate. Above this concentration citrate caused repression of the enzymes of glucose metabolism (Table 2.7). The authors stated that their results demonstrated that the specific activities of glucose enzymes can be increased either by increasing the glucose concentration or by lowering the citrate concentration. These findings were in accord with the regulation of glucose enzymes by induction with glucose and repression by citrate or its metabolic products.

In batch cultures of *Pseudomonas ovalis* growing on acetate (assimilated via the glyoxylate cycle), the addition of small amounts of succinate completely repressed synthesis of isocitrate lyase. However, addition of succinate (39 mM) to carbon-limited cultures in the chemostat ($D = 0{\cdot}54\ \text{hr}^{-1}$) gave only partial repression (Data of Kornberg, Herbert and Tempest quoted by Tempest, 1970). In contrast, the addition of 15 mM succinate to ammonia-limited cultures with acetate as the carbon source produced a 30-fold reduction in the specific activity of isocitrate lyase. It therefore seems to be the case that under ammonia limitation the level of succinate in the culture, although less than 1·0 mM, was sufficient to cause repression of acetate assimilation.

These few examples indicate how the use of mixed substrates provides a powerful tool for the study of enzyme regulation. Results of experiments concerning other growth-limiting nutrients should also prove of great value in understanding enzyme control.

2.3 DILUTION RATE

2.3.1 Types of Response to Varying the Dilution Rate

Enzyme synthesis can respond in five basic ways to changes in the dilution rate in continuous culture as shown in Figure 2.4. (1) Perhaps the simplest case is where there is no change in enzyme activity with dilution rate, (2) enzyme activity may increase with increasing dilution rate, (3) enzyme activity may decrease with increasing dilution rate, (4) there may be some particular dilution rate at which the enzyme activity passes through a maximum, (5) the enzyme activity may pass through a minimum of a given dilution rate.

Dean (1972) reviewed a number of papers on this subject according to the classification described above and Table 2.8 summarizes some of the data which he presented. Although in Table 2.8 there are no examples of a constant enzyme activity with changing dilution rate, two examples were mentioned by Dean. The lack of examples of constant enzyme activity may reflect the fact that the 'normal' situation is for activity to vary. There is, however, the possibility that such examples have been considered a 'nil' response by some workers and consequently have not been fully reported. Holstrom (1968) reported that the activity of streptokinase produced by a *Streptococcus* strain

Table 2.8 Effect of Dilution Rate (D) on the Activity of Enzymes. [*Reproduced with permission from A. C. R. Dean (1972). In* 'Environmental Control of Cell Synthesis and Function', *A. C. R. Dean, S. J. Pirt and D. W. Tempest (eds). Academic Press, London.*]

Enzyme	*Organism*	D (h^{-1})	*Rate-limiting nutrient*	*Carbon source*
	Activity passes through a maximum value at a given dilution rate			
α-Glucosidase	K. aerogenes	0·70	NH_4	Maltose
β-Galactosidase	K. aerogenes	0·25	C	Lactose
	E. coli B6B2	0·80	C	Lactose
	E. coli B6	0·70	C	Lactose
	E. coli B	0·80–1·20	C	Maltose[a,b]
Acid phosphatase	K. aerogenes	0·40	C or K	Glucose
		0·40	C	Maltose
		0·60	SO_4^{2-}	Glucose
Pullulanase	K. aerogenes	0·25	C	Pullulan
Amidase	Ps. aeruginosa[c]	0·30	C	Acetamide
Lipase (extracellular)	Anaerovibrio lypolitica	0·08 0·135	C	Glycerol
	Activity passes through a minimum value at a given dilution rate			
Glucose dehydrogenase	Ps. aeruginosa	0·17	$NH_4^+{}_+$	Citrate
Glucose-6-phosphate dehydrogenase	Ps. aeruginosa	0·25	NH_{4+}	Glucose-citrate
6-Phosphogluconate dehydrogenase	Ps. aeruginosa	0·20	NH_4	Glucose-citrate
Hexokinase	Candida utilis	0·40	C	Sucrose
	Activity increases as D is increased			
β-Galactosidase	K. aerogenes		K^+, NH_4^+, Mg^{2+}, SO_4^{2-} or PO_4^{3-}	Lactose

Enzyme	Organism	D (h^{-1})	Rate-limiting nutrient	Carbon source
Histidase	K. aerogenes LC1[d]		SO_4^{2-} or l-arginine	Histidine
Urease	Hydrogenomonas H16		C	Fructose
Glutamate dehydrogenase	E. coli		C	Glucose
	Aspergillus nidulans		C	Glucose
Glutamate-oxalate transaminase	E. coli		C	Glucose
Hexokinase[e]	Aspergillus nidulans		C	Glucose
Isocitrate dehydrogenase	Candida utilis		C	Sucrose
	Activity decreases as D increases			
β-Galactosidase	K. aerogenes		C	Maltose[a]
	K. aerogenes		K^+ or NH_4^+	Glycerol[a]
Acid phosphate	E. coli 308		C	Succinate[a]
	K. aerogenes		C	Sucrose
Hexokinase	E. coli		C	Glucose
Glucose-6-phosphate dehydrogenase	Candida utilis		C	Sucrose
Ornithine carbamyl transferase	Ps. fluorescens		C	Citrate

[a] TMG as inducer.
[b] Second stage of two-stage system; all other results were obtained in single-stage chemostats.
[c] Activity of constitutive mutant decreased with increasing D.
[d] Auxotroph with two blocks in arginine pathway.
[e] Fructose diphosphate aldolase and glucose-6-phosphate dehydrogenase activities also increased and to a lesser and greater extent respectively than hexokinase.

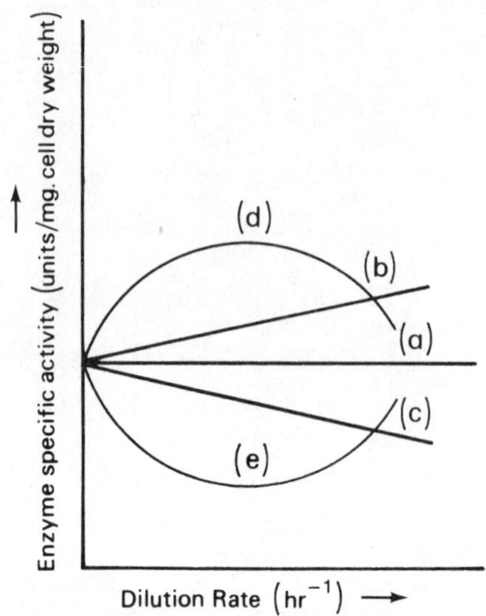

FIG. 2.4 Possible effects of dilution rate on enzyme activity in chemostat cultures.

was constant between $D = 0{\cdot}4$ and $1{\cdot}0\ hr^{-1}$, although this represents a fairly narrow range of dilution rates. The culture was grown in complex medium in which the nature of the growth-limiting nutrient was not known. Dean (1971) reported that the activities of glucose-6-phosphate dehydrogenase and glucose dehydrogenase produced by *K. aerogenes* were constant over the range $D = 0{\cdot}3$ to $1{\cdot}0$ in glucose-limited culture. However, reducing the dilution rate below $0{\cdot}3\ hr^{-1}$ resulted in a fall in glucose dehydrogenase and a corresponding increase in glucose-6-phosphate dehydrogenase activity. It would however be surprising if enzyme activities did remain constant since changes in dilution rate result in so many changes in cell composition and morphology (Herbert, 1961).

Before discussing the possible mechanisms for enzyme regulation with changing growth rate it would be useful to consider some examples in addition to those given in Table 2.8.

The behaviour of cariogenic streptococci has been studied by use of continuous culture particularly in relation to the production of enzymes associated with dental decay and adherence of organisms to the teeth. It has been found that dental plaque contains a high concentration of various carbohydrate polymers produced from sucrose and it was suggested that such polymers might trap bacteria and thus allow them to remain close to the tooth surface for extended periods (Gibbons, 1968). One enzyme which catalyses the formation of such polymers is glycosyl transferase and production of this enzyme has been studied. Carlsson and Elander (1973) grew *Streptococcus sanguis* in chemostat cultures using glucose limitation with a complex medium. They observed a progressive fall in the glycosyl transferase activity with increasing dilution rate. Activity was highest at $D = 0{\cdot}025\ hr^{-1}$ and was reduced to 1/5th of this when the dilution rate was $0{\cdot}33\ hr^{-1}$. It was perhaps significant that the reduced enzyme activity coincided with an increase in the level of unutilized glucose in the culture supernatant. When a chemically defined medium was used, also glucose limited, the glycosyl transferase activity was about 1/8th of that in the complex medium, but again activity decreased with increasing dilution rate. The authors reported that addition of cyclic AMP to the culture did not increase the glycosyl transferase activity. However, cyclic AMP was added to a slow growing (D = 0·08 hr^{-1}) culture in the defined medium where enzyme activity was near its maximum for that growth medium and where clearly the potential enzyme level was well below that in complex medium. It would be of interest to examine the effect of cyclic AMP in fast growing cultures on complex medium where unutilized glucose is present. Ellwood, Hunter and Longyear (1974) also studied the production of glycosyl transferase in *Streptococcus mutans* strain Ingbritt, using a complex medium with glucose as the growth-limiting substrate. They too found that maximum enzyme production occurred at low dilution rates and that at the lowest dilution rate studied (D = 0·05 hr^{-1}) a pH of 6·5 was critical for maximum production. Production of phosphoenolpyruvatephosphotransferase by *Strep. mutans* in glucose-limited complex medium was examined by Hunter, Baird and Ellwood (1973) and, like the glycosyl transferase, production of this enzyme decreased with increasing dilution rate. It is of particular interest that formation of these enzymes, which may be involved in adherence and acid production in oral streptococci, should be greatest at growth rates in the chemostat which are of the same order as those for organisms growing in the mouth.

Callow, Capel and Evans (1970, 1971) reported that asparaginase production by *E. carotovora* was reduced when the dilution rate was increased from 0·15 to 0·28 hr^{-1}. They used a complex medium with glycerol as the carbon source and found that at the high dilution rate the level of glycerol exceeded that which could be completely utilized. The asparaginase levels fell from

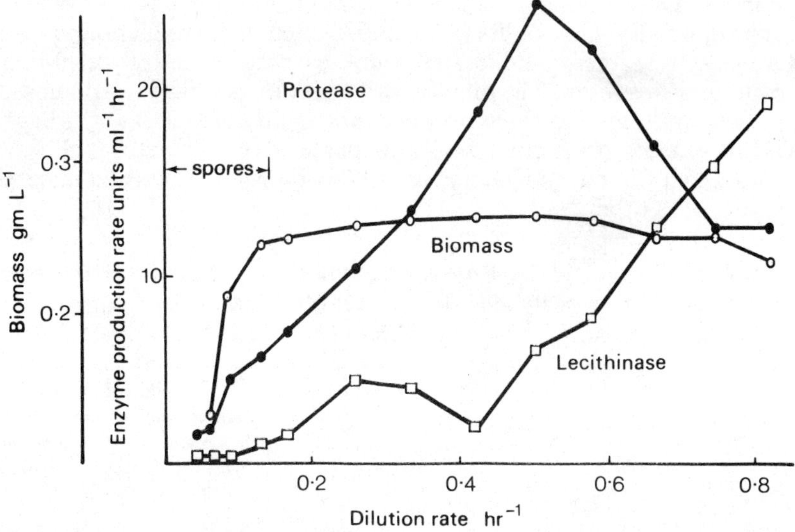

FIG. 2.5 Protease, lecithinase and biomass formation by *Clostridium b.fermentans* in continuous culture. [*Reproduced with permission from D. J. Cox and K. T. Holland (1976). In* 'Spore Research'.

197 units per ml to 2 units per ml with the increase in dilution rate. Asparaginase production was also examined by Liu and Zajuc (1973) using *Erwinia aroideae* growing on a complex medium with lactose. Again enzyme activity decreased with increasing growth rate.

Cox and Holland (1975) studied production of protease and lecithinase by *Clostridium bifermentans* in a single-stage continuous culture, under carbon-limited conditions in complex medium. As Figure 2.5 indicates, protease production showed a maximum at $D = 0{\cdot}5\ hr^{-1}$ while lecithinase production increased with increasing dilution rate. The authors suggested that protease production is controlled by repression in the presence of high amino acid to peptide ratios. It was further suggested that lecithinase may be attacked by the proteases and that this could partially explain the variations in activity.

Using a 2-stage continuous culture Jensen (1972) found that a low dilution rate was necessary for protease production from *B. subtilis.* There was little or no protease produced in the first stage ($D = 0{\cdot}16\ hr^{-1}$) but in the second stage ($D = 0{\cdot}06\ hr^{-1}$) protease equivalent to 60% of that obtained in batch culture was formed. Similarly Fabian (1969) obtained protease from *Bacillus pumilis* equivalent to 30% of batch production by using a 2-stage chemostat. Ryu, Lee and Thomas (1972) also used a 2-stage chemostat to produce 3-ketosteroid-delta-1-dehydrogenase from *Arthrobacter simplex.* They achieved a maximum cell yield in the first stage at a dilution rate of $0{\cdot}17\ hr^{-1}$ and then investigated the effect of dilution rate in the second stage on enzyme production. Dilution rates between 0·025 and $0{\cdot}065\ hr^{-1}$ were used and a peak in enzyme production was obtained at $D = 0{\cdot}045\ hr^{-1}$.

The production of both acid and neutral phosphatases varied with dilution rate (Bolton and Dean, 1972) as shown in Table 2.3. Phosphatase activity at both pH 4·6 and pH 7·1 fell with increasing dilution rate for all the growth-limiting nutrients used, except that activity at pH 7·1 under ammonia-limited conditions showed a maximum value at $D = 0{\cdot}7\ hr^{-1}$.

Lilley, Rowley and Bull (1973) investigated the production of β-1,3-glucanase by a thermophilic streptomyces species using glycerol-limited growth in a chemostat. The dilution rate was varied between 0·014 and 0·63 hr^{-1} and maximum enzyme production was attained at $D = 0{\cdot}2\ hr^{-1}$. A peak in enzyme production was also observed by Thurston (1973) who reported that Chondroitinase activity in *Proteus vulgaris* was at a maximum

Table 2.9 Effect of Dilution Rate and Growth-limiting Nutrient on the Specific Activity of UDP-N-acetylglucosamine dehydrogenase from *Achromobacter georgiopolitanum*. [*J. Melling, unpublished results.*]

Dilution rate (hr^{-1})	*Enzyme specific activity* (units/mg *dry weight*) *Growth-limiting nutrient*		
	Glucose	*Magnesium*	*Phosphorus*
0·05	nil	2·10	nil
0·10	nil	0·85	nil
0·12	3·5	0·32	nil
0·15	4·2	0·40	nil
0·2	4·0	nil	nil

at $D = 0{\cdot}19\ hr^{-1}$ over the range $D = 0{\cdot}08$ to $0{\cdot}4\ hr^{-1}$ when growing in chondroitin-limited culture. Production of UDP-N-acetylglucosamine dehydrogenase, an enzyme associated with cell envelope biosynthesis in *Achromobacter georgiopolitanum*, was examined (J. Melling, unpublished results). As the results in Table 2.9 show that enzyme specific activity responded differently to changes in dilution rate, depending upon the growth-limiting nutrient. The decrease in activity with increasing dilution rate in magnesium-limited culture and the complete absence of enzyme in phosphorus-limited conditions suggest that some switch in cell envelope biosynthesis may occur in response to the availability of magnesium. Although the wall composition has not so far been investigated in this organism, in *B. subtilis* a variation in composition was observed on changing from magnesium to phosphorus-limitation (Ellwood and Tempest, 1969) and other workers (Melling, Robinson and Ellwood, 1974; Robinson, Melling and Ellwood, 1974; Melling and Robinson, 1975) have shown that antibiotic sensitivity and cell envelope composition in the Gram-negative organism *P. aeruginosa* vary on changing from magnesium to phosphorus limitation. The increase in UDP-N-acetylglucosamine dehydrogenase with increasing dilution rate in glucose-limited culture (Table 2.9) may reflect the diminishing importance of the magnesium excess as the level of glucose in the culture rises.

Synthesis of three extracellular enzymes, phosphatase, staphylokinase and protease by *Staphylococcus aureus* was studied by Arvidson, Bjorkland, Eriksson and Holme (1976) and their results are shown in Table 2.10. Cultures were grown glycerol-limited using a complex medium. Phosphatase and protease were both apparently repressed by excess glycerol (and amino

Table 2.10 The Influence of Different Dilution Rates on the Growth of *Staphylococcus aureus* and on Exoenzyme Production. [*Reproduced with permission from S. Arvidson, A. Bjorkland, R. Eriksson and T. Holme. In* 'Continuous Culture: Applications and New Fields', *A. C. R. Dean, D. C. Ellwood, C. G. T. Evans and J. Melling (eds). Ellis Horwood, Chichester, 1976.*]

Dilution rate, h^{-1}	*0·07*	*0·15*	*0·3*	*0·4*	*0·6*
Bacterial dry weight, mg/ml	1·65	2·2	2·6	1·6	0·8
Y_{O_2}, g bacterial dry weight per mole O_2	17	30	30	44	40
Glycerol in medium, g/l	0	0	0	0	0·2
$Y_{glycerol}$, g bacterial dry weight per g glycerol	0·66	0·88	1·0	0·62	0·35
Phosphatase, units per mg bacterial dry weight	0·65	1·5	1·8	4·0	1·5
Staphylokinase, μg per mg bacterial dry weight	0·1	0·3	0·15	0·9	2·0
Protease I, μg per mg bacterial dry weight	0	0	0	1	0

acids) but staphylokinase was produced when these constituents were in excess.

2.3.2 Possible Regulatory Mechanisms Involved in the Influence of Dilution Rate on Enzyme Activity

Not all of the possible patterns of variations in enzyme activity mentioned above are as yet explicable. However, in at least two of the cases good evidence exists to account for the observed effects.

The progressive decline in enzyme activity with increasing dilution rate which has frequently been observed appears to be linked with the increase which occurs in the concentration of growth-limiting nutrient with increasing dilution rate (Herbert, Elsworth and Telling, 1956). In two cases of organisms constitutive for particular enzymes, a fall in enzyme activity with increasing dilution rate was observed. Using a mutant of *Pseudomonas aeruginosa* constitutive for amidase and growing on a minimal medium containing acetamide and succinate, it was observed (Clarke, Houldsworth and Lilly, 1968) that the amidase specific activity fell with increasing dilution rate. Production of β-galactosidase by a constitutive mutant of *E. coli* followed a similar pattern of declining enzyme activity with increasing dilution rate (Mandelstam, 1962). These results are entirely consistent with a progressively increasing degree of catabolite repression as the dilution rate increased. Further evidence for this increase in catabolite repression of an enzyme which did not require induction by the limiting nutrient came from the work of Smith and Dean (1972). By using the non-metabolizable inducer MTG they were able to maintain a non-constitutive strain of *K. aerogenes* fully induced for β-galactosidase over a range of dilution rates. Under carbon limited conditions they observed the expected fall in β-galactosidase activity with increasing dilution rate. However, a similar fall in enzyme activity in an ammonia-limited chemostat when the carbon source was always in excess was also seen. Furthermore the findings by Silver and Mateles (1969) and of Ricia, Necinova, Stejkalova and Fencl (1967), that in 2-stage continuous culture constitutive strains of *E. coli* and *K. aerogenes* as well as a strain of *E. coli* induced by MTG all showed a maximum enzyme activity with increasing dilution rate, indicate that the simple explanation of increasing catabolite repression of a constitutive enzyme does not hold in all cases.

The experiments of Clarke, Houldsworth and Lilly (1968) provided an interpretation of the maximum enzyme activities which occur with increasing dilution rate. They suggested that, as the dilution rate increased at the lower end of the range, the effective concentration of the inducer at the inducer binding site increased. However, as the dilution rate was increased still further the concentration of catabolite repressor molecules also increased, until a dilution rate was reached after which catabolite pression became dominant over induction. The wild-type *Ps. aeruginosa* exhibited a typical bell-shaped curve of amidase activity with increasing dilution rate in contrast to the decline in amidase activity shown by the constitutive mutant. The authors also suggested that if such induction-repression controls existed for this enzyme it was likely that different curves would be obtained for those mutants which were known by their growth on plates and in batch culture to have altered amidase regulation and (as described above) this proved to be so.

In the case of enzyme activities which increased with increasing dilution rate no firm explanation is yet available for these results. However, Jensen and Neidhardt (1969) observed an increase in histidase activity of *K. aerogenes* as the dilution rate was raised. They concluded that catabolite repression was a controlling factor in histidase biosynthesis and suggested that such repression might be overcome by an increase in those pathways which convert repressors to inactive forms. Thus by analogy with the decline in enzyme activity with increasing dilution rate and the appearance of maximum values for enzyme production, so minimum values might result from maximum repressor levels followed by an increase in the metabolism of such compounds. Such suggestions are still speculative and more work is needed, perhaps by looking for mutants with altered abilities to metabolize repressor substances, to elucidate this problem.

2.3.3 The Concept of Relative Growth Rate

It has been tacitly accepted for some time that direct comparisons of the effects of dilution rate on, for example, cell yields of cultures growing on different substrates or at different temperatures may be open to question and that it may be more meaningful to consider the dilution (growth) rate as a proportion of the maximum growth rate of the organism under the particular conditions (C. G. T. Evans, 1971, personal communication). That the use of such relative, rather than actual, growth rates in assessing results of continuous culture experiments was valid was demonstrated by Tempest (1975, VIth International Continuous Culture Symposium discussion) and the topic has been more fully discussed by him since then (Tempest, 1976).

According to Monod (1950) the specific growth rate of an organism (μ) is related to its maximum growth rate (μ_{max}) according to the equation

$$\mu = \mu_{max} \cdot \frac{s}{K_s + s}$$

where s is the steady state concentration of growth-limiting nutrient and K_s a saturation constant that is numerically equal to the concentration of growth-limiting nutrient, which gives $\mu = \frac{1}{2}\,\mu_{max}$. Hence

$$s = K_s \cdot \frac{\mu}{\mu_{max} - \mu}$$

or in the chemostat

$$s = K_s \cdot \frac{D}{\mu_{max} - D}$$

If a change is now imposed on the culture, such as a reduction in growth temperature or in a culture limited by phosphorus, the incorporation of a 'poorer' carbon source, giving a lower maximum growth rate, then because of the reduction in μ_{max} the steady-state concentration of the limiting nutrient must change (assuming K_s remains constant). Thus to maintain the same steady-state concentration of growth-limiting nutrient after a change affecting μ_{max}, the dilution rate must be changed in proportion to the change imposed on μ_{max}. This may perhaps be clarified by a hypothetical example.

Consider a culture growing in steady state at 35 °C with $D = 0{\cdot}4\ \text{hr}^{-1}$ such that

$$\mu_{max} = 1{\cdot}0\ hr^{-1}$$

then
$$s = K_s \cdot \frac{0{\cdot}4}{1{\cdot}0 - 0{\cdot}4}$$

$$\therefore s = 0{\cdot}66 \times K_s$$

If the growth temperature is then reduced to 25 °C we can assume from the Arrhenius equation that for this 10 °C reduction $\mu_{max\ 25°} = \frac{1}{2}\,\mu_{max\ 35°}$ and therefore maintaining $D = 0.4$ gives a value for the concentration of limiting nutrient

$$s = \frac{0{\cdot}4}{0{\cdot}5 - 0{\cdot}4}$$

$$\therefore s = 4\,K_s$$

Thus the level of growth-limiting nutrient increases and to maintain the former level, the dilution rate must be reduced to $\frac{1}{2}$ its previous value. However, the *relative* growth rate value μ/μ_{max} remains unchanged.

Since in many of the examples considered in this chapter the level of growth-limiting nutrient is critical for the regulation of cell metabolism it is clear that the concept of relative growth rate is important. Although no data are yet available on enzyme levels with respect to relative growth rate, the following examples indicate how this concept may be applied. Tempest (1976) also made the point that it has only been assumed that factors which cause μ_{max} to vary do not also affect K_s, which is difficult to measure. Thus it is important to examine some experimental data to see how well this accords with the theory.

Tempest (1976) cited two examples of data which could be re-interpreted using relative, rather than actual, growth rates. Hill, Drozd and Postgate (1972) showed that at any fixed dilution rate the steady-state biomass concentration of *Azotobacter chroococcum* grown sulphur-limited was greater when ammonia was the nitrogen source than when the organisms were fixing nitrogen. The same authors also showed the maximum growth rate of this organism growing on molecular nitrogen to be only 80% of that when ammonia was supplied. If cell yields for the two nitrogen sources were plotted against relative, instead of actual, growth rate the two curves were superimposable.

Tempest and Hunter (1965) showed that at a fixed dilution rate of 0·2 hr^{-1} the RNA content of *K. aerogenes* varied markedly with the growth temperature. However, μ_{max} also varied with temperature and Tempest (1976) concluded that the RNA contents of organisms growing at the same relative growth rate were approximately equivalent irrespective of temperature.

Further examples concerning the growth of *Bacillus stearothermophilus* were given by Evans (1976). This organism has an unusually high minimum growth rate and it was found that in complex medium with sucrose as limiting carbon source using a dilution rate of 0·2 hr^{-1} growth occurred at 50 °C but not at 65 °C presumably because at 50 °C 0·2 hr^{-1} represents a relative growth rate above the minimum but at 65 °C 0·2 hr^{-1} expressed as a relative growth rate is below the minimum. This suggestion was further supported since growth would occur at 65 °C using $D = 0{\cdot}2\ \text{hr}^{-1}$ if glycerol was the carbon source because μ_{max} is less with glycerol than sucrose and thus $0{\cdot}2\ \text{hr}^{-1}$ is, in terms of relative growth rate, above the minimum value. For similar reasons growth at 65 °C using sucrose did occur with $D = 0{\cdot}2$ in a defined salts medium since μ_{max} in the minimal medium is less than μ_{max} in the complex medium.

In an elegant theoretical treatment (Herbert, 1976) has proposed the symbol ϕ to represent the relative growth rate thus:

$$\phi = \frac{\mu}{\mu_{max}}$$

and for continuous culture applications the relative dilution rate was defined as:

$$W = \frac{D}{\mu_{max}}$$

and of course in a continuous culture in the steady state

$$W = \phi$$

He has further pointed out that while μ and μ_{max} have the dimensions of $(\text{time})^{-1}$ their ratio ϕ (and W) is dimensionless i.e. is a pure number. In the same way Herbert (1976) pointed out that three more dimensionless variables could be defined:

relative substrate concentration $\sigma = \dfrac{s}{K_s}$

relative cell concentration $\chi = \dfrac{x}{\gamma K_s}$

relative time $\tau = \mu_{max}\, t$

and thus the fundamental equations for continuous culture could be re-written using dimensionless quantities.

The main relevance of these dimensionless quantities, however, Herbert considered to reside in the physiological concepts they express. The significance of the relative growth rate (ϕ) has already been discussed and similarly the relative substrate concentration (σ) is a convenient measure of the extent to which the cell is saturated with the growth-limiting nutrient and since σ determines ϕ it is the more fundamental quantity although difficult to measure owing to the low values of K_s.

Considered in the context of regulation of enzyme biosynthesis in response to particular nutrient limitations σ assumes considerable importance.

2.4 MUTANT SELECTION

The chemostat can provide highly selective environments with respect to various physiological characteristics of micro-organisms including the formation of particular enzymes. Indeed, in one of the early papers concerning the development of the theory of continuous culture Novick and Szilard (1950) considered the problem of selection for and against mutants in the chemostat.

In the absence of some specific selection pressure continued growth in continuous culture will favour those micro-organisms which are most efficient metabolically. This means that those organisms which have eliminated physiological systems which are not required for growth in a particular system (Zamenhof and Eichhorn, 1967).

Most of the strains which are used to produce enzymes and various metabolites are metabolically inefficient in that they grossly overproduce materials for which they have no need. Such strains therefore are highly susceptible to take-over by spontaneously arising mutants in which overproduction has been reduced or eliminated, unless selective pressures can be applied to favour the hyper-producer. It may be the difficulty of maintaining such pressures over long periods that has restricted the use of continuous culture for production of microbial products. Conversely, where suitable conditions have been used there has often been evidence of an improvement in yield of the required product on prolonged cultivation. The chemostat also provides a system in which the evolutionary process can be studied (Francis and Hansche, 1972).

An organism can respond in various ways to a stress involving a particular enzyme and an examination of these will exemplify the potential of the chemostat as a tool for selection and maintenance of organisms with desired characteristics.

Situations have been examined where an increase in the catalytic activity of an enzyme or affinity for its substrate would give an advantage and in several cases the response to both conditions has been increased enzyme production rather than a change in K_m.

Novick and Horiuchi (1961) observed that when *E. coli* was grown in lactose-limited culture there was initially a selection for constitutive mutants which could produce β-galactosidase at the maximum capacity even without an inducer. On continued cultivation they isolated organisms which produced four times as much β-galactosidase as the constitutive strains and which they called hyper-producers. It was demonstrated by Horiuchi, Tomizawa and Novick (1962) that the enzyme produced by the hyper strains was identical to the original. Thus the increased enzyme activity indicated that the β-galactosidase represented about 25% of the cell protein. A similar take-over by a strain with increased ability to take up tryptophane was observed when *E. coli* was grown in a chemostat with tryptophane as the limiting nutrient. The mutant was found to exhibit faster growth on low tryptophane concentrations than the parent (Novick and Szilard, 1950).

A rapid development of hyper strains of *E. coli* with respect to β-galactosidase resulted from growth under carbon-limited conditions using mixed substrates of glucose and lactose (Silver and Mateles, 1969). They argued that there was greater selective pressure in the presence of the two substrates since the ability to use lactose freed the bacteria from the restraint imposed by glucose limitation. Hyperactivity for β-galactosidase was also obtained by Smith and Dean (1972) in *K. aerogenes* grown lactose-limited. These authors suggested that in the presence of two substrates the specific growth rate would increase from D to $D+X$ where X is the increment due to lactose utilization. Thus lactose utilizers should outgrow the rest at a rate proportional to $(D+X)/D$ which clearly increases as D is reduced.

In a study of enzyme evolution Rigby, Burleigh and Hartley (1974) grew *K. aerogenes* using xylitol as the limiting nutrient and examined the activity of ribitol dehydrogenase. This enzyme has a side specificity for xylitol with an apparent K_m for xylitol of 1 M compared with a K_m of 1 mM for ribitol. It was likely therefore that the activity of ribitol dehydrogenase would limit growth on xylitol. It was found that both spontaneous mutants and those produced by UV irradiation or treatment with nitrosoguanidine exhibited superproduction of the enzyme rather than a change in specificity. This increased production was achieved by gene duplication and the authors concluded that, at least for this enzyme, the development of altered specificity required gene duplication as a first step, followed by mutation of one copy to a 'silent' gene which produces an inactive product due to incorrect folding. The product of the 'silent' gene can then become heterogeneous in the population since it is no longer subject to natural selection. Later, mutations accumulated during the silent phase may be expressed after reversion of the original lesion and the modified enzymes subjected to natural selection.

Conversely Betz, Brown, Smyth and Clarke (1974), using a conventional plating technique, were able to isolate amidases with altered substrate specificities. Changes in substrate specificity were also observed in the case of an acid phosphatase of *S. cerevisiae* (Francis and Hansche, 1973). Cultures were limited by β-glycerophosphate at an unfavourably high pH (6·0) whereas the optimum pH of the original enzyme was 4·2. After about 290

Table 2.11 Acid Phosphatase Activities of *Saccharomyces cerevisiae* over a Range of pH with β-glycerophosphate (BGP) as Substrate. S288C was the Original Strain. M4 was isolated after some 290 Generations in Chemostat Culture at pH 6·0 with BGP as Limiting Nutrient. [*Reproduced with permission from J. C. Francis and P. E. Hansche (1973).* Genetics **74,** *259–265.*]

pH	*Activity** (μmole/min/cell) × 10^{-10} S288C	M4
3·6	22·86±0·16	34·34±0·11
3·8	22·85±0·15	34·80±0·24
4·0	23·72±0·11	35·15±0·16
4·2	25·84±0·09	37·51±0·29
4·4	24·03±0·25	41·19±0·14
4·6	22·53±0·10	45·98±0·15
4·8	21·62±0·18	50·02±0·21
5·0	19·84±0·12	53·23±0·26
5·2	16·87±0·10	55·81±0·19
5·4	11·43±0·13	49·44±0·13
5·6	8·09±0·08	41·74±0·18
6·0†	6·72±0·19	26·60±0·23

* Estimates are reported with 95% confidence intervals.

† Estimates at pH 6 were made with a different buffer; see Francis and Hansche (1972)—*Materials and Methods.*

generations an increase in the population density occurred and it was found that the mutant strain, which had taken over contained an acid phosphatase with a pH optimum of 5·2 and showed a four-fold increase in activity at pH 6·0, compared with the original (Table 2.11).

Selection for an altered enzyme was also detected by Downie and Garland (1973) who grew *Candida utilis* in continuous culture under copper-limited conditions. After about 7 days at a dilution rate of 0·2 hr^{-1} they found that the dry weight had increased several-fold and the culture at that stage exhibited a decreased requirement for copper. Further investigation revealed that the variant was able to by-pass the cytochrome oxidase by utilizing an alternative oxidase which communicated with the respiratory chain at about the level of cytochrome b. In non copper-limited culture the normal cytochrome oxidase again functioned, but the alternative enzyme was not lost.

The type of mutant selected (whether K_m or V_{max}) in continuous culture may be influenced by the growth conditions. Thus where a substrate, even a poor one, is in excess and a high D is used this could favour the selection of V_{max} mutants, i.e. exhibiting increased enzyme production. Conversely at low dilution rates and with low substrate concentrations K_m mutants may be favoured. It would be of interest to examine this situation in practice.

2.5 SOME CONCLUSIONS

The technique of continuous culture offers advantages for studying the regulation of enzyme synthesis in relation to microbial physiology because it permits well-defined variations in the bacterial environment to be made and reproduced from occasion to occasion. This also makes the technique useful for improving production of microbial enzymes by selecting appropriate conditions for over-production of the required material; although process conditions may still favour batch culture.

Growth under such conditions is also available in the selection of (mutant) organisms with certain desired characteristics and this area, which has only begun to be explored, may prove to be one of the most valuable aspects of continuous culture. Indeed one can envisage the selection and maintenance of particular strains in a chemostat which then provide starter cultures for subsequent production cultures.

Chemostat cultures may also be used to simulate natural environments and the capacity of various ecosystems to cope with certain effluents is an aspect worthy of further consideration. In the medical field, cultures grown in the chemostat may be useful in providing defined inocula for evaluating antibiotics and disinfectants. In particular the ability to model, even partially, the *in vivo* situation could prove valuable in the selection and assessment of chemotherapeutic agents for treating microbial infections. It should almost certainly be an advance on the use of batch culture and the lack of definition implicit therein.

Although it is not yet possible to produce ‘a microbe for all seasons’ the technique of continuous culture undoubtedly is of value towards achieving this end.

REFERENCES

Acevedo, F. and Cooney, C. L. (1973), *Biotechnol. Bioeng.*, **15**, 493–503.

Aiba, S., Shoda, M. and Nagatani, M. (1968), *Biotechnol. Bioeng.*, **10**, 845–864.

Akbar, M. D., Rickard, P. A. D. and Moss, F. J. (1974), *Biotechnol. Bioeng.*, **16**, 455–474.

Andrews, J. F. (1968), *Biotechnol. Bioeng.*, **10**, 707–723.

Arvidson, S., Bjorkland, A., Eriksson, R. and Holme, T. (1976), in *Continuous Culture: Applications and New Fields.* Ed. by A. C. R. Dean, D. C. Ellwood, C. G. T. Evans and J. Melling. Ellis Horwood, Chichester.

Asthana, H., Humphrey, A. E. and Moritz, V. (1971), *Biotechnol. Bioeng.*, **13**, 923–929.

Betz, J. L., Brown, P. R., Smyth, M. J. and Clarke, P. H. (1974), *Nature*, **247**, 261–264.

Bolton, P. G. and Dean, A. C. R. (1972), *Biochem. J.*, **127**, 87–96.

Brown, C. M. (1976), in *Continuous Culture: Applications and New Fields.* Ed. by A. C. R. Dean, D. C. Ellwood, C. G. T. Evans and J. Melling. Ellis Horwood, Chichester.

Brown, C. M. and Johnson, B. (1970), *J. gen. Microbiol.*, **64**, 279–287.

Brown, C. M., Macdonald-Brown, D. S. and Meers, J. L. (1974), *Adv. Micro. Physiol.*, **11**, 1–52.

Brown, C. M., Macdonald-Brown, D. S. and Stanley, S. O. (1973), *Antonie van Leeuwenhoek*, **39**, 89–98.

Burn, V. J., Turner, P. R. and Brown, C. M. (1974), *Antonie van Leeuwenhoek*, **40**, 93–102.

Callow, D. S., Capel, B. J. and Evans, C. G. T. (1970), British Patent Appln. No. 54277/70.

Callow, D. S., Capel, B. J. and Evans, C. G. T. (1971), British Patent Appln. No. 8662/71.

Carlsson, J. and Elander, B. (1973), *Caries Res.*, **7**, 89–101.

Clarke, P. H., Houldsworth, M. A. and Lilly, M. D. (1968), *J. gen. Microbiol.*, **51**, 225–234.

Dawson, P. S. S. (1972a), in 'Fermentation Technology Today'. *Proc. IVth Internat. Fermentation Symp.* Ed. by G. Terui. Soc. Ferment. Technol. Japan, 121–128.

Dawson, P. S. S. (1972b), *J. appl. Chem. Biotechnol.*, **22**, 79–103.

Dean, A. C. R. (1972), *J. appl. Chem. Biotechnol.*, **22**, 245–259.

Dean, A. C. R., Gray, S. C., Hope, G. C. and Johnson, I. M. (1972), *Biochem. J.*, **126**, 15p–16p.

Dean, A. C. R. and Moss, D. A. (1971), *Biochem. Pharmac.*, **20**, 1–14.

Dean, A. C. R. and Rogers, P. J. (1969), *Nature*, **221**, 969–971.

Demain, A. L. (1972), *J. appl. Chem. Biotechnol.*, **22**, 345–362.

Downie, J. A. and Garland, P. B. (1973), *Biochem. J.*, **134**, 1051–1061.

Edwards, V. H. (1970), *Biotechnol. Bioeng.*, **12**, 679–712.

Ellwood, D. C., Hunter, J. R. and Longyear, V. M. C. (1974), *Archs oral Biol.*, **19**, 659–664.

Ellwood, D. C. and Tempest, D. W. (1969), *Biochem. J.*, **111**, 1–5.

Eisenberg, R. C., Butters, S. J., Quay, S. C. and Friedman, S. B. (1974), *J. Bacteriol.*, **120**, 147–153.

Evans, C. G. T. (1976), in *Continuous Culture: Applications and New Fields.* Ed. by A. C. R. Dean, D. C. Ellwood, C. G. T. Evans and J. Melling. Ellis Horwood, Chichester.

Evans, C. G. T. and Kite, S. (1964), in *Continuous Cultivation of Micro-organisms.* Ed. by I. Malek, K. Beran and J. Hospodka. Czech. Acad. Sci. Prague, 299–309.

Fabian, J. (1970), *Folia Microbiol.*, **15**, 160–168.

Fahien, L. A., Lin-Yu, J. H., Smith, S. E. and Happy, J. M. (1971), *J. Biol. Chem.*, **246**, 7241–7249.

Francis, J. C. and Hansche, P. E. (1972), *Genetics*, **70**, 59–73.

Francis, J. C. and Hansche, P. E. (1973), *Genetics*, **74**, 259–265.

Gibbons, R. J. (1968), *Caries Res.*, **2**, 164–171.

Gorin, L. and Maas, W. K. (1957), *Biochim. Biophys. Acta*, **25**, 208–209.

Griffith, C. J. and Carlsson, J. (1974), *J. gen. Microbiol.*, **82**, 253–260.

Harder, W., Attwood, M. M. and Quale, J. R. (1973), *J. gen. Microbiol.*, **78**, 155–163.

Heineken, F. G. and O'Connor, R. J. (1972), *J. gen. Microbiol.*, **73**, 35–44.

Herbert, D. (1958), in 'Recent Progress in Microbiology'. *Symp. 7th Int. Congr. for Microbiology*, 381–396.

Herbert, D. (1961), in 'Microbial Reaction to Environment'. *11th Symposium Soc. Gen. Microbiol.*, 391–416.

Herbert, D. (1976), in *Continuous Culture: Applications and New Fields.* Ed. by A. C. R. Dean, D. C. Ellwood, C. G. T. Evans and J. Melling. Ellis Horwood, Chichester.

Herbert, D., Elsworth, R. and Telling, R. C. (1956), *J. gen. Microbiol.*, **14**, 601–622.

Herbert, D. and Phipps, P. J. (1974), *Soc. Gen. Microbiol. Proc.*, **1**(3), 70.

Holstrom, B. (1968), *Appl. Microbiol.*, **16**, 73–77.

Horiuchi, T., Tomizawa, J. I. and Novick, A. (1962), *Biochim. Biophys. Acta*, **55**, 152–163.

Hovinu, J. C., Conway, R. A. and Ganze, C. W. (1973), *Pollut. Control Fed.*, **45**, 71–84.

Jacob, F. and Monod, J. (1961), *J. Molec. Biol.*, **3**, 318–356.

Jensen, D. E. (1972), *Biotechnol. Bioeng.*, **14**, 647–662.

Jensen, D. E. and Neidhardt, F. C. (1969), *J. Bacteriol.*, **98**, 131–142.

Jones, G. L. and Carrington, E. G. (1972), *J. Appl. Bacteriol.*, **35**, 395–404.

Katz, R., Kilpatrick, L. and Chance, B. (1971), *Eur. J. Biochem.*, **21**, 301–307.

Kavanagh, B. M. and Cole, J. A. (1976), in *Continuous Culture: Applications and New Fields.* Ed. by A. C. R. Dean, D. C. Ellwood, C. G. T. Evans and J. Melling. Ellis Horwood, Chichester.

Levine, D. N. and Cooney, C. L. (1973), *Appl. Microbiol.*, **26**, 982–990.

Light, P. A. (1972), *J. appl. Chem. Biotechnol.*, **22**, 509–526.

Lilley, G., Rowley, B. I. and Bull, A. T. (1973), *J. appl. Chem. Biotechnol.*, **23**, 167–168.

Liu, F. S. and Zajic, J. E. (1973), *Appl. Microbiol.*, **25**, 92–96.

Magasanik, B. (1961), *Symp. Quant. Biol.*, **26**, 249–256.

Magasanik, B., Prival, M. J., Brenchley, J. E., Tyler, B. M., De Leo, A. B., Streicher, S. L., Bender, R. A. and Paris, C. G. (1974), in *Current Topics*

in Cellular Regulation. Ed. by B. L. Horecker and E. R. Stadtman. Academic Press, New York, **8**, 119.
Mandelstam, J. (1962), *Biochem. J.*, **82**, 489–493.
Mateles, R. I. (1975), *Proc. 1st Int. Congress of IAMS Vol. 5.* Ed. by Science Council of Japan, Tokyo, 30–36.
Mateles, R. I., Chian, S. K. and Silver, R. (1967), in *Microbial Physiology and Continuous Culture.* Ed. by E. O. Powell, C. G. T. Evans, R. E. Strange and D. W. Tempest. H.M.S.O. London, 233–239.
Meers, J. L. (1973), *J. appl. Chem. Biotechnol.*, **23**, 163.
Melling, J. and Brown, M. R. W. (1975), in *Resistance of Pseudomonas aeruginosa.* Ed. by M. R. W. Brown. Wiley, London, 35–70.
Melling, J. and Robinson, A. (1975), *Soc. Gen. Microbiol. Proc.*, **3**, 68–69.
Melling, J., Robinson, A. and Ellwood, D. C. (1974), *Soc. Gen. Microbiol. Proc.*, **1**, 61.
Monod, J. (1942), *Recherches sur la Croissance des Cultures Bacteriennes*, Herman, Paris.
Monod, J. (1950), *Ann. Inst. Pasteur.*, **79**, 390–410.
Monod, J., Changeux, J. P. and Jacob, F. (1963), *J. Molec. Biol.*, **6**, 306–329.
Neijssel, O. M., Hueting, S., Crabbendam, K. J. and Tempest, D. W. (1975), *Arch. Microbiol.*, **104**, 83–87.
Ng, F. M.-W. and Dawes, E. A. (1973), *Biochem. J.*, **132**, 129–140.
Novick, A. and Horiuchi, T. (1961), *Symp. Quant. Biol.*, **26**, 239–245.
Novick, A. and Szilard, L. (1950), *Proc. Nat. Acad. Sci.*, **36**, 708–719.
Paigen, K. and Williams, B. (1970), *Adv. Microbiol. Physiol.*, **4**, 251–324.
Perlman, R. L., de Crombrugge, B. and Pastan, I. (1969), *Nature*, **223**, 810–812.
Poole, R. K. and Haddock, B. A. (1975), *Biochem. J.*, **152**, 537–546.
Powell, E. O. (1956), *J. gen. Microbiol.*, **15**, 492–511.
Powell, E. O. (1958), *J. gen. Microbiol.*, **18**, 259–268.
Powell, E. O. (1972), in *Environmental Control of Cell Synthesis and Function.* Ed. by A. C. R. Dean, S. J. Pirt and D. W. Tempest. Academic Press, London, 71–78.
Prival, M. J., Brenchley, J. E. and Magasanik, B. (1974), *Proc. Nat. Acad. Sci.*, U.S.A., **71**, 225–229.
Rieche, A., Hilgetag, G., Martin, A. and Lorenz, M. (1964), in *Continuous Cultivation of Microorganisms.* Ed. by I. Malek, K. Beran and J. Hospodka. Czech. Acad. Sci., Prague, 293–298.
Rigby, P. W. J., Burleigh, B. D. and Hartley, B. S. (1974), *Nature*, **251**, 200–204.
Robinson, A. and Tempest, D. W. (1973), *J. gen. Microbiol.*, **78**, 361–370.
Robinson, A., Melling, J. and Ellwood, D. C. (1974), *Soc. Gen. Microbiol. Proc.*, **1**, 61.
Robinson, G. A., Butcher, R. W. and Sutherland, E. W. (1971), *Cyclic AMP*, Academic Press, New York, 423–440.
Ryu, D. Y., Lee, B. K. and Thoma, R. W. (1972), *Proc. 4th Int. Ferm. Symp. Fermentation Technol. Today.* Ed. by G. Terui. Society of Fermentation Technol., Japan, 129–133.
Sayer, P. D. (1968), *Appl. Microbiol.*, **16**, 326–329.
Silver, R. and Mateles, R. I. (1969), *J. Bacteriol.*, **97**, 535–543.

Smith, R. W. and Dean, A. C. R. (1972), *J. gen. Microbiol.*, **72**, 37–47.

Snedecor, B. and Cooney, C. L. (1974), *Appl. Microbiol.*, **27**, 1112–1117.

Standing, C. N., Fredrickson, A. G. and Tsuchiya, H. M. (1972), *Appl. Microbiol.*, **23**, 354–359.

Takada, N., Sawada, H., Sawada, S., Amano, Y. and Terui, G. (1972), *Abstracts 4th Int. Fermentation Symp.* 19–25 March, Kyoto.

Tempest, D. W. (1970), *Adv. Microbiol. Physiol.*, **4**, 223–250.

Tempest, P. W., Meers, J. L. and Brown, C. M. (1970), *Biochem. J.*, **117**, 405–407.

Tempest, D. W. (1976), in *Continuous Culture: Applications and New Fields.* Ed. by A. C. R. Dean, D. C. Ellwood, C. G. T. Evans and J. Melling. Ellis Horwood, Chichester.

Thurston, C. F. (1973). *J. gen Microbiol.*, **75**, xviii.

Torriani, A. (1960), *Biochim. Biophys. Acta*, **38**, 460–469.

Tubb, R. S. and Postgate, J. R. (1973), *J. gen. Microbiol.*, **79**, 103–117.

Tyler, B., De Leo, A. B. and Bagasanik, B. (1973), *J. Biol. Chem.*, **248**, 4334–4344.

Van Uden, N. (1967), *Arch. Mikrobiol.*, **58**, 145–154.

Whiting, P. H., Midgley, M. and Dawes, E. A. (1976), *J. gen Microbiol.*, **92**, 304–310.

Zamenhof, S. and Eichhorn, H. H. (1967), *Nature*, **216**, 456–458.

Chapter **3**

Foam Separation of Biological Materials

A. THOMAS and M. A. WINKLER, Department of Chemical Engineering, University of Surrey, Guildford, Surrey

SYMBOLS

A	surface area
A_s	specific surface
a	empirical constant
a_i	activity of species 'i'
B, b	empirical constants
C	concentration of bulk solution
C_f	foamate concentration
C_o	feed concentration
C_r	residual or raffinate concentration
c	empirical parameter
d	bubble diameter
E	enrichment ratio
F	volume fraction of gas in foam
G	gas volume flow-rate
G_f	Gibbs free energy
G_s	superficial gas velocity
g	gravitational acceleration factor
H	foam column height
k	constant of proportionality
L	volume of liquid drainage
L_o	original volume of liquid
m	empirical constant
n	number of capillaries per unit cross-sectional area of foam
n_m	number of moles of gas
P	pressure, atmospheric pressure
P_{av}	average pressure in foam
P_c	total length of capillaries per unit vertical length of foam
R	universal gas constant
S	entropy
S_f	empirical constant
T	absolute temperature
t	time
u	average liquid velocity in a capillary
V	volume
V_b	volume of solution entrained by a single bubble
V_f	volume flow-rate of foamate
V_0	volume flow-rate of feed
V_r	volume flow-rate of residual liquid or raffinate
V_l	volume entrained by a single bubble on formation
v_f	volume of foamate
w	empirical exponent
Z	empirical parameter
Γ	surface excess
Γ_i	surface excess in equilibrium with solution 'i'
γ	surface tension

δ	bubble wall thickness	ε_f	foam density
ε	volume fraction of liquid	μ	viscosity
ε_c	fractional cross-sectional area for liquid flow	ρ	density

3.1 INTRODUCTION

Foam separation is a method of separating components of a solution by utilizing differences in surface activity. It is one of a number of related techniques known collectively as 'adsorptive bubble separation processes' which make use of the property of certain species of substances to concentrate at a liquid surface. This concentration effect may be relative to the species itself or to other species present in the system. Such substances are said to be 'surface-active'. The liquid surface at which surface-active species are concentrated are in most processes gas-liquid interfaces, as in foam separation: some processes make use of the interfaces between two liquid phases and are known as 'emulsion' or 'droplet' separation processes. The principle can also be used for separating non-surface-active species by attachment to surface-active substances known as 'collectors'.

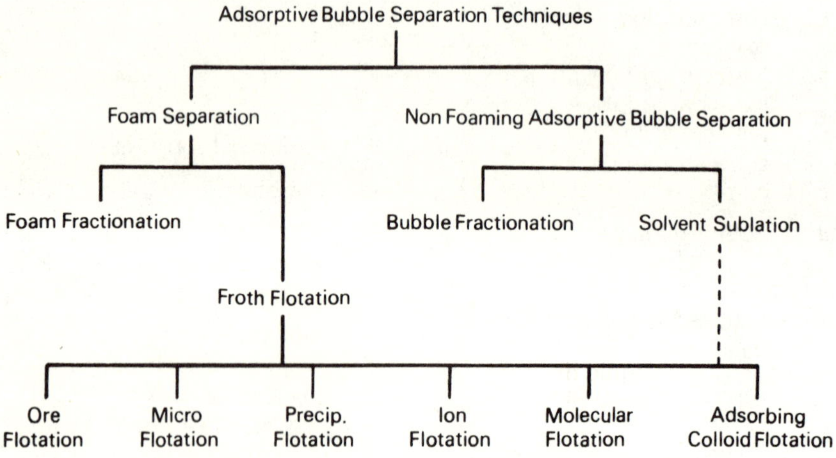

FIG. 3.1 Adsorptive bubble separation techniques.

The complete range of adsorptive bubble separation techniques is illustrated diagrammatically in Figure 3.1 (Karger, Grieves, Lemlich, Rubin and Sebba, 1967): of these various techniques, foam separation processes are of most relevance to the separation of biological materials. They can be considered in two subdivisions, 'foam fractionation' and 'froth flotation'. Foam fractionation gives separation or concentration of a species from a homogeneous solution, while froth flotation is used to separate or concentrate particulate matter. The distinction between the two becomes somewhat blurred when considering colloidal systems or those containing the very large molecules characteristic of biological material.

3.2 BASIC PRINCIPLES

3.2.1 The Surface Excess

The concentrated layer formed by surface-active substances at a surface is called the 'surface excess'. The surface excess arises because a surface-active molecule can assume at the surface an orientation more stable than when it is completely surrounded by liquid in the bulk of the solution. A molecule of a surface-active substance contains groups with differing affinities for the solvent: in aqueous solutions, these groups are either hydrophilic or hydrophobic. Within the bulk of the solution, the lack of affinity of hydrophobic groups for the solvent water causes the water to assume an ordered, low-entropy structure: this implies a high-energy, low-stability arrangement. At the water surface, the molecule can align itself with hydrophilic groups orientated towards the aqueous phase and the hydrophobic groups towards the non-aqueous phase. The non-aqueous phase may be a gas, as in foam separation, or a non-polar solvent, as in emulsions. This configuration is more stable, and the free energy of stabilization has been quoted as 4 kcal/mole for every (unspecified) hydrophobic side-chain of a dissolved protein removed from an aqueous environment to a non-polar environment (Charm, 1972). Since protein molecules contain large numbers of both hydrophilic and hydrophobic groups, many are consequently surface-active. The configurational modifications resulting from changes in the protein environment will affect its surface activity, and not all proteins expose hydrophobic groups to the surrounding solution: lysozyme, for example has most of its hydrophobic groups concealed within the interior of the molecule.

The existence of the surface itself is stabilized by the presence of the surface excess, when the free energy reduction in the surface excess balances the surface potential energy. If it does not, the tendency of the surface to assume minimum area will cause coalescence or collapse of a foam.

When a surface is created in a solution containing surface-active material, the surface-active molecules diffusing to the surface will tend to remain there, forming the surface excess and increasing their concentration near the surface. Eventually, an equilibrium is reached where the chemical potential of the high concentration of the low-energy configuration equals that of the lower concentration of the high-energy configuration in the bulk of the solution, and a concentration gradient will exist near the surface. The time taken to reach equilibrium at a static surface has been measured and can be several minutes or even hours for some proteins (Evans, Mitchell, Mussellwhite and Irons, 1970). The surface may then 'age' as the molecules at the surface re-orientate themselves or, as can happen with protein molecules, undergo distortion or unfolding caused by the stresses between the hydrophilic and hydrophobic groups. The difference in surface tension, termed 'surface pressure', between the solution and the surface excess causes stresses which can result in distortion of large molecules. This has led to the widely held belief that proteins are denatured by foaming, which is a possible reason why foam separation processes have not been more widely adopted for protein purification. While it is certainly true for some proteins, many enzymes have been studied which have not been found to undergo denaturation at surfaces, and, indeed, many enzymes function naturally at interfaces such as cell membranes.

3.2.2 Mechanism of Separation

Foam separation achieves its effect by the creation of a large gas-liquid interfacial area in the form of gas bubbles in the liquid. The liquid around the bubbles becomes enriched in the surface-active species, with respect to the bulk of the solution, and this enriched liquid is carried with the bubbles as they rise through the solution to form a foam at the top of the liquid. If the foam is removed and allowed to collapse, the resultant solution consists entirely of enriched liquid.

The formation of the surface excess is very rapid, as the bubbles effectively 'sweep' the solution, bringing interface to the surface-active solute, rather than allowing equilibrium to be attained by molecular diffusion to the interface, as in the static-surface situation. If the bubbles arriving at the top of the liquid are insufficiently stable to form a foam, or if foam is not removed and is allowed to collapse back into the liquid, surface-active species are swept from the bottom to the top of the bulk solution, setting up a concentration gradient. This effect is known as 'bubble fractionation' (Lemlich, 1972) and can be considered as a separation process in its own right (Dorman and Lemlich, 1965) in a non-foaming system, and can make an important contribution to the efficacy of a foaming separation system, where the enriched liquid is removed in the form of foam.

3.2.3 Process Outline

Foam separation is carried out in a vessel commonly called a 'foaming cell'. Foam is produced by forming bubbles in the solution held in the cell by one of several alternative methods.

The usual method of forming bubbles is to force gas into the liquid through sparger nozzles or through perforated- or sintered-plate distributors at the bottom of the cell. The gas flow-rate is one of the most significant operating parameters in a foam separation system: to allow for the effect of cell diameter, it is most conveniently expressed as 'superficial gas velocity'. The gas volume flow-rate determines principally the number of bubbles passing through the liquid in a given time, the superficial gas velocity expresses the number per unit cross-sectional area of the cell. The diameter of the bubbles formed by the sparger or distributor is important, as this determines the interfacial area presented by the gas to the solution and also the velocity of rise of the bubbles through the solution, and thus the contact time of bubbles and solution.

The foam at the liquid surface may, if it is sufficiently stable, be allowed to build up to a considerable height above the top of the liquid. This enables the foam to be easily removed from the cell, but, more importantly, further enrichment of surface-active species occurs in the foam column, so that a high foam column is associated with a concentrated product. The reasons for this effect are discussed later. The height of the foam column attainable depends on the concentration of surface-active material in the solution and on the superficial gas velocity: a high column even of unstable foam can be obtained if the gas velocity is high enough.

The three process variables most significant in determining the degree of enrichment of surface-active material gained in the process are effectively interdependent: (1) the concentration of surface-active material in the solution, (2) the superficial gas velocity, (3) the height of the foam column from which the foam is removed. In the separation of proteins, the environmental factors which affect protein structure and thus its surface activity will also be important: pH, ionic strength, presence of 'organic' solvents, and in the surface excess layer, the surface pressure. Within the 'normal' ambient range, temperature has not been found very significant.

There are other means of forming bubbles, sometimes used in water-treatment processes. Foam may be produced by mechanical agitation, though this has been shown to be a definite cause of protein denaturation (London *et al.*, 1953). Gas can be dissolved in the solution under pressure so that a foam is formed when the pressure is released, as in bottled beer (and indeed, beer foam has been shown to contain 73% protein (Charm, 1972)); in industrial processes, it is most conveniently used in continuous systems, where gas is dissolved in a pressurized vessel, the pressure being released when the liquid enters the foaming cell (Rosen, 1971). Gas bubbles can also be produced by electrolysis of aqueous solutions, as in the waste-treatment process known as 'electroflotation'.

3.2.4 Modes of Operation

A foam separation cell can be operated in several different modes, as illustrated in Figures 3.2 to 3.6 (Lemlich, 1968). In the simple batch mode (Figure 3.2), a volume of solution is made to foam and the foam is removed from the foam column as it reaches a specified height. The collected foam is allowed to collapse, to give the product known as the 'foamate' (unfortunately, some authors have confusingly used the term 'foamate' to refer to the residual liquid left in the cell after foaming). As in all such batch operations, the composition of both the foamate and the residual solution change throughout the foaming process, and as in batch distillation, a 'cut' of desired composition can be taken from the foamate stream. The volume of solution falls as liquid is removed in the foam: a possible variant on this mode could be to add just sufficient fresh solution to keep the volume in the foaming cell constant, but there is no report of this procedure having been used.

In the simple continuous mode (Figure 3.3), solution is separated as a once-through procedure, by adding fresh solution to the bulk liquid continuously, and removing not only the foamate, but a proportion of the bulk liquid as well. In this, as in the other continuous modes, a situation is reached where the composition of the foamate and of the bulk solution in the cell stabilize at constant values.

A variation on this simple continuous mode is the 'stripper' mode (Figure 3.4), where the stream of fresh solution enters the system in the foam column rather than the bulk solution. As it falls through the foam, the fresh liquid strips material from the foam back into the cell. The other continuous modes involve returning a proportion of the foamate to the cell. When this is done in the simple continuous mode, it is called the 'enricher' mode (Figure 3.5); on the stripper mode, it is known as the 'combined' mode (Figure 3.6).

The use of the complex modes does give an improvement in the efficacy of separation, but with biological materials, structural damage can result from repeated subjection to foaming and foam-breaking. This was certainly found with foam fractionation of LDH (Charm, 1972), where returning

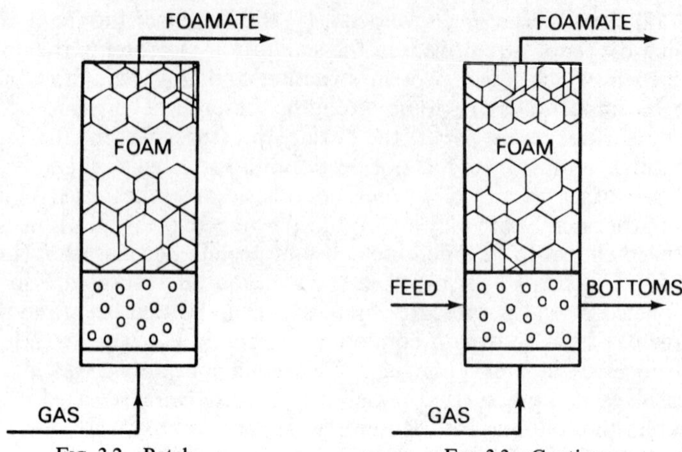

FIG. 3.2 Batch. FIG. 3.3 Continuous.

Modes of simple operation, foam fractionation.

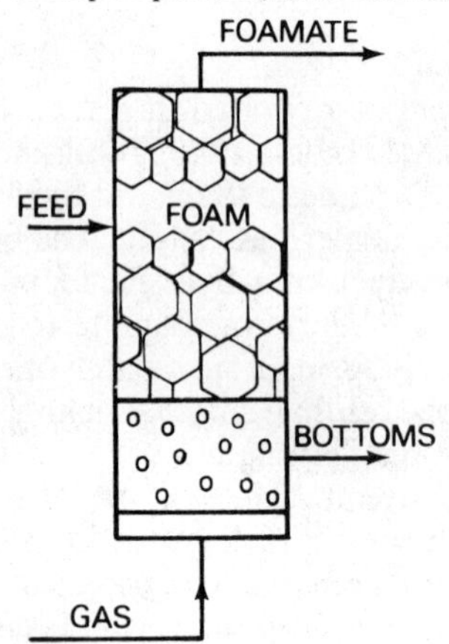

FIG. 3.4 Continuous stripper.

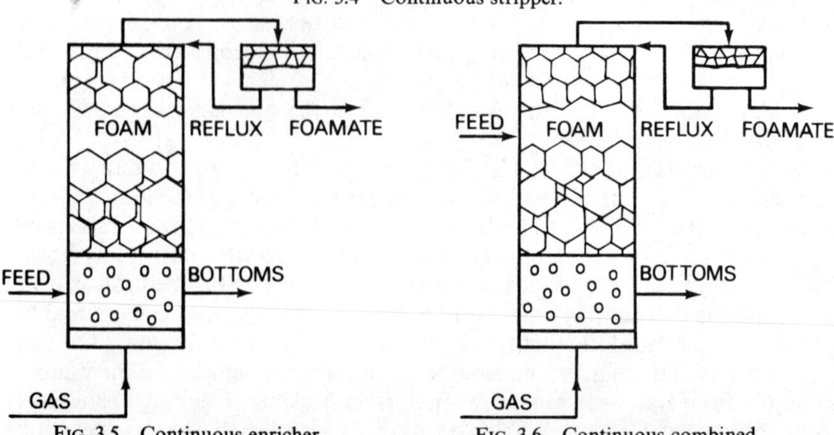

FIG. 3.5 Continuous enricher. FIG. 3.6 Continuous combined.

Various other modes, continuous foam fractionation.

foamate to the foaming cell increased the enrichment ratio to fourfold, compared with twofold in the simple batch operation, but severe loss of enzyme activity was experienced. Thus the complex operating modes are useful only where damage to the surface-active molecules is not important, as in, for example, recovering protein for use in cattle fodder.

3.3 PREVIOUS WORK ON FOAM FRACTIONATION

3.3.1 Early Work on Fractionation of Biological Materials

Foam fractionation was first used to separate albumen from potato and beet juices (Ostwald and Siehr, 1937), using a very simple apparatus. Gonadotrophic hormones in the urine of pregnant women have been concentrated by foaming (Courrier and Dognon, 1939) and carboxymethyl cellulose has been fractionated by foaming according to molecular weight and degree of methylation (Schütz, 1942). Several groups of workers used foam fractionation to separate and purify enzymes (Ostwald and Mischke, 1940; Schütz, Bader and Stacey, 1944; Dognon and Gougerot, 1947, for example) and a notable achievement was the separation of rennin and pepsin (Andrews and Schütz, 1945). These proteolytic enzymes were so difficult to separate that for a long time they were thought to be identical. When a purified pepsin solution was foamed at a pH value in their normal working range (pH 1·2 to 2·0), the pepsin activity was concentrated in the foam, while the rennin activity was left in the residual solution.

In work by the same school on the purification of bile (Bader and Schütz, 1946), the bile salts, sodium glycocholate and taurocholate, were separated by enrichment in foam. This is an interesting illustration of the analogy between emulsions containing a large interfacial area between two liquid phases, and foams with gas-liquid interfaces. To perform their natural function of emulsification and 'solubilization' of lipids (Mahler and Cordes, 1966), involving the creation and stabilization of interfacial area between lipid and aqueous phases, the bile salts must be surface-active and could therefore be expected to concentrate in foam. Although there was no standardization or assaying of the solutions involved in this work on foam fractionation, useful indications of the effects of certain process parameters were given.

Foam was produced by passing gas into the bulk solution through a single nozzle, and it was noticed that low gas bubbling rates produced a richer foamate and that changes in solution pH resulted in changes in the efficacy of separation. It was acknowledged that the difficulty of distinguishing between individual proteins made enzymes the most promising subjects for investigating the performance of fractionation systems; but, evidently as a result of fears regarding denaturation, work on foam fractionation continued, for a time, mostly in the non-biological field.

3.3.2 Non-biological Applications of Foam Fractionation

The rapid increase in the use of non-biodegradable synthetic detergents gave rise to problems in sewage treatment plants, where aeration units created large volumes of foam by unintentional foam fractionation. Since these detergents were not broken down microbiologically in the sewage treatment process, they contaminated natural waters into which sewage-plant effluent was discharged and caused further foaming problems. The difficulty was overcome initially by carrying out deliberate, controlled foam fractionation in specially constructed plants to remove the surface-active material (Stephan, 1965), but there still remained the problem of disposing of the concentrated foamate. The product of one foam fractionation plant in California was dumped in a desert area! The long-term answer came with the introduction of biodegradable synthetic detergents. This did not entirely remove the need for the foam fractionation plants, which are now in use as a final polishing stage to remove residual traces of detergent before discharge: the foamate is returned to one of the biological treatment stages (Jenkins, Scherfig and Eckhoff, 1972).

Considerable expertise in the engineering of foam fractionation units was built up during the development of foaming units for detergent stripping, and long-term studies of the behaviour of surfactant foaming systems have led to the development of a generalized equation describing the rate of detergent stripping from a solution (Grieves and Bhattacharya, 1970); for a solution detergent concentration C_r, and a gas volume flow-rate G, the detergent concentration in the foamate, C_f, and the volume of foamate, v_f, are described by

$$\frac{d}{dt}(C_f \cdot v_f) = S_f \cdot (G)^m \cdot (C_r)^w \tag{3.1}$$

where t is time and S_f, m and w are constants.

However, performance data of plant operating on detergent foams are not directly applicable to systems involving biological materials, as there is considerable difference between the behaviour of detergent foam and those of biological surface-active materials such as saponins and albumens (Shih and Lemlich, 1971), due, at least in part, to the considerable difference in molecular size.

More recent applications of non-biological foam fractionation have been in removing metallic ions and complexes from solution by attachment to surface-active collectors. Copper and zinc have been extracted from sea-water in this way (Zwierzkowski and Medrzyka, 1973), and 150-fold enrichment in the removal of mercury^{II}-ions from solution has been reported (Miller and Sullivan, 1971). Weakly surface-active dyes, such as crystal violet chloride which are insufficiently surface-active to foam, have been concentrated by bubble fractionation (Harper and Lemlich, 1966; Shah and Lemlich, 1970).

3.3.3 Recent Work on Foam Fractionation of Biological Materials

In work on the foam fractionation of enzyme preparations (London, Cohen and Hudson, 1954; London and Hudson, 1953), results of earlier workers discussed above were confirmed, and useful indications of the effects of certain important process parameters on both purification and denaturation of the enzymes were given, although from a limited experimental programme. Catalase (pork liver), urease (jackbean) and acid phosphatase (prostate) were purified individually from solutions of their respective preparations, and catalase and urease were separated from a mixed preparation of the two. Urease was concentrated in the foam, while catalase and acid phosphatase tended to remain in the residual solution and appeared in the foam only towards the end of a batch run. Protein damage was associated with the making and breaking of unstable foam, such as that produced by mechanical agitation, rather than with foams produced by the introduction of gas.

Of the two gases used for foam production, the less stable foams produced with carbon dioxide caused some denaturation, while the 'wet', stable foams from nitrogen caused no such damage. In addition to confirming the findings of Schütz (1946) that the enzymes gave the highest purification and recovery from solutions at pH values near their respective iso-electric points, it was concluded that purification and recovery both depend on protein concentration and that the highest recoveries were obtained at particular concentrations characteristic of the protein. The values quoted were 0·08% for urease, 0·037% for catalase and 0·01% for acid phosphatase, with enrichment ratios of 9·7, 2·3 and 15 respectively.

It was also reported that purification was reduced by the addition of sodium sulphate and of ethanol, and increased by addition of non-ionic detergent. Foams with small bubbles gave high recoveries, as did large diameter foaming cells; there was an optimum foam column height for each cell diameter, and the depth of solution in the cell was also thought to be significant. Temperature was not found to affect the behaviour of the system significantly in the range 10 to 30 °C, and as the gas flow-rate was not strictly controlled, no conclusion could be drawn with respect to that parameter.

Foams causing bloat in cattle have been investigated in a qualitative manner with foaming experiments on lucerne solutions (Buckingham, 1970) and on bovine salivary secretions with their associated protozoal proteins (Jones and Lyttleton, 1972). The protozoal proteins were found to act as foam stabilizers, undergoing surface denaturation at pH 5·8 and giving maximum foam persistence in the pH range 5·5 to 6·5.

Foam fractionation of a number of different substances, including lipoid material, was studied with a more fundamental and quantitative approach in a simple foam fractionation system with a constant height of foam column and without removal of the foamate (Karger, 1966). The mechanism proposed by which enrichment in the foam was achieved depended on internal reflux, as foamate from collapsed foam drains back through the foam column, where any surface-active solute would be preferentially adsorbed at the interface in the foam. Recoveries of surface-active ionic salts varied from 40% immediately on foam formation to 100% after 1 hour. Recovery of vegetable lecithin (C_{12} to C_{18}) of 70% in the foam after 2 hours was achieved using an anionic surfactant, sodium lauryl sulphate, in acid conditions.

The recovery of protein from liquors containing dissolved or precipitated casein and haemoglobin by foaming was investigated as a comparison with the activated sludge process (Iibuchi, Yano and Torikata, 1974). Recovery of precipitated protein was more efficient than that of dissolved protein and compared favourably with the activated sludge process. Temperature had little effect in the range 10 to 40 °C, but the addition of salts gave a marked improvement, particularly in acid conditions.

Experiments with β-lactoglobulin solutions showed that surface tension decreases and foam stability increases with increasing concentration of added salts (Lauwers, 1964). Bovine serum albumen (BSA) previously added to sea-water has been successfully removed at concentrations as low as 5 microgrammes per litre (Wallace and Wilson, 1969): from consideration of the rate-limiting processes occurring in foam separation at such low concentrations, it was concluded that the operation was strongly dependent on gas bubble diameter and gas flow-rate. BSA solutions were also used to confirm London, Cohen and Hudson's inferences by foaming with air and with the concentrated foamate returned to the bulk solution (Schnepf and Gaden, 1959). Enrichment ratios up to 20 times were reported, with the concentration of BSA solutions as low as 20 mg/litre, and no denaturation was detected. Measurements of the surface tension of BSA solutions at different concentrations showed a very steep drop in surface tension with increasing BSA concentration in dilute solutions, levelling off to a barely detectable rate of decrease at high concentrations. From this, it was predicted and observed that enrichment ratio increases sharply with decreasing BSA concentration. Maximum enrichment was obtained at the iso-electric pH of the protein, and the presence of a non-surface-active polymer, dextran, of molecular weight similar to that of BSA had no effect on the enrichment of BSA and neither was itself enriched. It was inferred that the gas used to produce foam would affect the degree of protein denaturation caused by the process, with carbon dioxide, air and 'inert' gases in decreasing order of denaturing power. As foam fractionation gives increasing enrichment with decreasing protein concentration, a conclusion drawn was that the process is best suited to preliminary concentration of very dilute protein solutions before further concentration by other techniques.

The same apparatus was used at the New England Enzyme Centre to investigate the foam fractionation of solutions of catalase and β-amylase preparations and the separation of the enzymes from mixed solutions of the two (Charm, Morningstar, Matteo and Paltiel, 1966). Measurements of surface tension of the enzyme preparations at different concentrations in water and in ammonium sulphate solution showed that there are upper and lower threshold concentrations between which surface tension falls evenly with increasing protein concentration, and is virtually constant outside the thresholds. Both thresholds are lower, for their respective enzymes, in ammonium sulphate solution than in water. Between the concentration thresholds, separation efficiency is most influenced by solution pH and is highest near, but not at, the iso-electric pH of the individual enzyme. Successful separation of both enzymes, individually from their preparations and from each other was effected under conditions predicted from the surface tension-concentration data. After exhausting one batch of enzyme solution, further separation was obtained by increasing the concentration of ammonium sulphate and foaming again. Losses in enzyme activity were about 15% for catalase and 5% for the amylase. Foaming trials were carried out on six other enzymes: cellulase, D-amino acid oxidase (hog kidney) tripeptide synthetase (bakers' yeast) and aryl pyruvate keto-enol tautomerase (*Candida tropicalis*) showed negligible loss of activity, while ADH (human liver) showed 5–20% activity loss with 5-fold purification, and malic dehydrogenase (chicken heart) showed a 25% loss of activity. Later work on LDH (chick heart) showed no separation of the enzyme on foaming, but its specific activity in the residual solution was doubled as a result of the removal of non-active protein into the foam (Charm, 1972).

Streptokinase was separated by foaming from a crude streptococcal culture filtrate (Holmström, 1968), but was unusual in that at the iso-electric pH of the enzyme, pH 5, enrichment was poor and the deactivation was 40 to 50%. At pH 6·5 however, enrichment of 3·4 times was obtained with 80% recovery, and the least inactivation, of 6%, was found at pH 7. The explanation advanced for this was that streptokinase is sensitive to low pH levels, which caused the inactivation at its iso-electric pH, and the high recovery at pH levels considerably higher than its i.e.p. was due to its adsorption on to other protein in the solution which was then concentrated in the foam.

3.3.4 Discussion of Previous Work

Some general inferences can be drawn from the above work regarding the effects of the process parameters on the efficacy of separation achieved in foam fractionation, and to a certain extent on the risks of protein denaturation, and have been confirmed by the present authors' recent experimental results.

3.3.4.1 Parameters Affecting Separation. The variables found to have most effect on separation, as indicated by the enrichment ratios obtained, are the protein concentration of the solution in which the foam is formed, the gas flow-rate (or superficial gas velocity), the height of foam column from which foamate is taken and the aqueous environment of the protein. Other variables shown to have some effect are the gas bubble size, the depth of solution through which bubbles rise in the foaming cell and the foaming cell diameter.

The enrichment ratio increases with decreasing concentration of protein in the bulk solution. The rate of increase of enrichment rises rapidly for very dilute solutions, down to a limiting concentration below which there is insufficient protein for formation of the surface excess layers and thus a foam. This is in accord with the implications of the fall in surface tension with increasing protein concentration. In a batch foaming operation, the concentration of surface-active material in the solution falls as the material is separated into the foam, with a consequent rise in enrichment ratio throughout the run. The rise in this ratio results from change in foamate concentration as well as the fall in solution concentration.

Enrichment ratios have also been found to increase with decreasing gas flow-rates, down to a limiting flow-rate where gas bubbles arrive too infrequently at the top of the solution for a foam to form. When a stable foam is formed, the higher the foam column is allowed to rise before foam is removed to give the foamate, the greater will be the enrichment; this is due to drainage and/or coalescence of the surface excess layers comprising the foam. Several mechanisms have been advanced to account for this effect, and they are discussed in more detail in Section 3.5.

The enrichment of any substance by foaming depends in the first instance on its surface activity; with proteins, the environmental factors affecting molecular configuration will also affect the separability by foaming. The most significant of these factors is the pH of the protein solution, and proteins in general being most susceptible to foam separation near their iso-electric pH. Other factors known to influence protein configuration, such as ionic strength and the presence of organic solvents have been shown to affect the process, although temperature is not very significant within the 'normal' ambient range.

3.3.4.2 Denaturation. One of the principal objections to foam fractionation as a protein separation process is the possibility of denaturation. The work discussed above has shown that there is considerable variation among different proteins in their susceptibility to denaturation by foaming. The risk is reduced by using inert gas to produce foam, by using low gas flow-rates, by avoiding mechanical agitation, by producing stable, wet foams and by using a once-through rather than a recycle process. In addition to the enzymes shown to resist denaturation (Charm, 1966), the present authors have found that by following these precepts that no loss of activity, within the accuracy of the assay procedures used, results from foam fractionation of solutions of preparations of acid- and alkaline phosphatases, catalase, β-amylase, hexokinase, phosphoglucoisomerase, alcohol dehydrogenase or β-galactosidase; not all of these were separated into the foamate fraction.

3.4 FROTH FLOTATION

When gas bubbles are formed in a slurry of solid particles, certain species of particle may become preferentially attached to bubbles and be carried into a froth at the top of the liquid. The froth can then be skimmed off and the particles allowed to settle out of the collapsed froth. Particles which are not themselves sufficiently surface-active to attach directly to bubbles can be collected by attachment to surface-active substances, known as 'collectors'. For example, particles which are electrically charged can be collected by attachment to an ionic surfactant of opposite charge. To enable the froth to exist long enough to be skimmed, a foam stabilizer may be added to the slurry, and many substances serve as both collectors and foam stabilizers. Other substances can be added which inhibit collection of unwanted species. The process was developed originally in the mineral dressing industry for separating valuable ore particles from gangue, bubbles being produced by mechanical agitation, and is still undergoing a lengthy conversion from an art to a science. The standard introductory text on the process is by Gaudin (1957) who was also one of the first to extend the principle to the separation of micro-organisms. Theoretical aspects of the process are comprehensively covered by Derjauguin (1961), and a text by Klassen and Mokrusov (1963) pays particular attention to work reported in the Soviet bloc countries.

3.4.1 Froth Flotation of Biological Materials

Loss of micro-organisms in foam from aerated cultures has long been a problem (Black, MacDonald and Gerhardt, 1958; Boyles and Lincoln, 1958) and much work has gone into the prevention of foaming in fermentation. Use of the effect was first made in removing bacterial cells from salt solutions (Dognon, 1941). Solid particles were removed from surface water by foaming (Hopper and McCowen, 1952) including 99% of bacteria and all cysts of

Endamoeba histolytica; in another water-treatment study (Moore and Bryant, 1954) bacteria were removed by foaming using quaternary ammonium salts as foaming agents.

Spores of *Bacillus anthracis* were separated from a well autolysed culture containing casein hydrolysate, by foaming with air, giving a clean concentrate of spores free of vegetative cells and cell debris, and with enrichment ratios of up to 19 times (Boyles and Lincoln, 1958). With a second foam concentration stage, enrichment ratios averaging 40 times could be obtained, and differences in the amount of cell debris in the spore concentrates was noticed between rhizoid and mucoid strains of the bacillus. Spore concentrates with enrichment ratios up to 8·6 were obtained from autolysed cultures of *Bacillus subtilis* var. *niger* when the culture pH was raised to 11·5 and a re-suspension of the spores after washing was easily reconcentrated by foaming. In the same work, concentration of cells of *Serratia marcescens* with enrichments of about 8 times was obtained, and with cells of *Bacillus suis*, separation of different strains according to their cell capsular material was achieved, but attempts to concentrate cells of *Pasteurella tularensis* were unsuccessful.

Separation of spores from an autolysed culture of *Bacillus cereus* was achieved by frothing (Black, MacDonald and Gerhardt, 1958) and vegetative cells were removed from a culture of *B. subtilis* var. *niger* (the same strain as used by Boyles and Lincoln) in a medium containing casein hydrolysate by frothing with air (Gaudin, Mular and O'Connor, 1960a). In an extension of this work, the same authors (1960b) showed that separability depended on the age of the spores, and while secondary amines acted as preferential collectors for spores, carboxylic acids collected vegetative cells and their debris; some inferences as to the surface structure of spores and vegetative cells were drawn from this. Separation was also affected by the solution pH according to the collector in use. Similar results were reported from work on *Escherichia coli*, with BSA as a foam stabilizer (Gaudin, Davis and Bangs, 1962a; 1962b) where spore age and solution pH were also found to be significant factors.

A considerable volume of work on the foam separation of bacteria has been carried out by Grieves and various co-workers. In experiments with a washed suspension of *E. coli* (Grieves and Wang, 1966), a cationic surfactant (EHDA-Br) was used as both collector and foam stabilizer in a simple batch system in which foam was produced with nitrogen. The bacterial cell concentration in the residual liquid was found to follow an exponential decay pattern with time, and enrichment ratios between 10 and 10^6 were obtained. As with foam fractionation, the very high enrichment ratios are achieved towards the end of the experiment when the residual solution is very depleted. Surfactant is apparently bound to the bacterial cells, as indicated by a discrepancy in the surfactant mass balance before and after foaming which increases with the quantity of bacteria used. In a continuation of the work (Bretz, Wang and Grieves, 1966), dosing of surfactant to the system in increments during foaming was found to give higher recoveries than by using the same total quantity of surfactant added at the start of the experiment. A simple linear relationship between final and initial concentrations of bacteria was proposed, but was found to hold at only one concentration of surfactant.

In another series of experiments on several different microbial species, the presence of inorganic salts was found to decrease the recovery of *Bacillus subtilis, Pseudomonas fluorescens, Serratia marcescens, Bacillus cereus, P. vulgaris* and *E. coli*, and salts with divalent ions had more effect than monovalent salts (Grieves and Wang, 1967a; 1967b). Except for *E. coli*, the presence of salts increased the volume of foamate produced: the variation of foam volume produced with solutions containing different species of micro-organism or the same micro-organism with different concentrations of surfactant has not been satisfactorily explained. There was no strong correlation between the Gram type of the species and the enrichment achieved or the foam volume produced, and no mechanism for the adsorption of surfactant by bacterial cells has been advanced.

Workers in Japan have separated a parent sake-yeast from its non-foaming mutant (Ouichi and Akuyama, 1971; Ouichi and Nunokawa, 1973), and differences in bubble adsorption between glucose-grown yeast and the same strain grown with a hydrocarbon carbon source showed that the cell walls of the hydrocarbon-grown strain were seven times more hydrophobic than those of the glucose-grown strain (Miyazu and Yano, 1974). Recovery of algae from cultures using froth flotation techniques have been investigated (Rubin *et al.*, 1966); the process used was characterized by the use of very low gas flow-rates, of collectors and foaming agents and flocculating agents such as alum. Experiments in a range of pH values between pH 4 and 8 showed that pH was critical, and optimal recovery of 90% was achieved at pH 7·2; the algae investigated were *Chlamydomonas reinhardii* and *Chlorella ellipsoida*. A similar system, using laurylamine or lauric acid as collectors gave reasonable recoveries of *Aerobacter aerogenes* without the use of a flocculating agent (Rubin, 1968). The process has also been widely applied to sludge thickening in sewage treatment plants (Jenkins *et al.*, 1972; Rosen, 1971).

3.5 THEORETICAL CONSIDERATIONS

The complexity of foaming systems makes them unsusceptible to mathematical analysis, and treatments starting from fundamentals have generally needed considerable inclusion of simplifying assumptions before reasonable correlation with experimental behaviour can be obtained. Foaming systems can conveniently be considered firstly in terms of the properties of foams themselves, and then in terms of the surface phenomena causing enrichment.

3.5.1 Foams

The importance of foam is not merely because it acts as a convenient means of collecting surface excess layers, but also because a considerable amount of further enrichment takes place within the foam as a result of liquid drainage and film coalescence. Models developed to describe foam properties are considered here firstly in terms of a static volume of foam, then with models accounting for drainage and then with consideration of the dynamic foam columns more usually encountered in foam separation systems.

3.5.1.1 Static Foam. An absolutely pure solvent will not foam, but minute quantities of surface-active material dissolved in it will allow foaming to occur (Nakagaki, 1957). In the same work, a thermodynamic theory was advanced to account for foam formation and foam stability in terms of foam volume ('foaminess') and foam life. This thermodynamic argument was said later to be incomplete, since it lacked terms to account for the effects of gravity and the pressure differences in the system (Ross, 1967). An equation of state was then developed (Ross, 1969) based on the classical work on the space structure of coalescing bubbles (Tait, 1885). For a number of moles of gas n_m, a volume of gas V at an absolute temperature T, the surface area A can be related to the outside atmospheric pressure P and the surface tension of the liquid γ. Tait's expression for coalescing spherical droplets

$$3P \cdot \Delta V + 4\gamma \cdot \Delta A = 0 \tag{3.2}$$

can be modified to apply to bubbles, which have two surfaces:

$$3P \cdot \Delta V + 2\gamma \cdot \Delta A = 0 \tag{3.3}$$

where V and A are the changes in bubble volume and surface area respectively. From these, Ross developed the expression

$$P \cdot \Delta V + \tfrac{2}{3}\gamma \cdot \Delta A = n_m \cdot R \cdot T \tag{3.4}$$

as a general law applicable to all foams, where R is the universal gas constant. Equation (3.4) can be rearranged in the form

$$P_{av} = \frac{n_m \cdot R \cdot T}{V} = P + \tfrac{2}{3}\gamma \frac{A}{V} \tag{3.5}$$

where P_{av} is the average pressure within the foam. This illustrates that, in agreement with thermodynamic principles, the area of surface per unit volume of foam is determined uniquely by the surface tension of the liquid and the average excess pressure in the foam relative to the atmosphere.

As was mentioned qualitatively in Section 3.2.1, the stability of a surface may be expressed in terms of free energy changes: for a one-component system that has sufficient surface for the surface energy to make a significant contribution to the total energy content, the Gibbs free energy, G_f, is given by

$$dG_f = V \cdot dP - S \cdot dT - \gamma dA \tag{3.6}$$

where S is entropy. At constant temperature and pressure,

$$\gamma = \left(\frac{\partial G_f}{\partial A}\right)_{P,\,T} \tag{3.7}$$

which on integration gives

$$\Delta G_f = \gamma \cdot \Delta A \tag{3.8}$$

where ΔG_f and ΔA are the changes in the Gibbs free energy and surface area at constant pressure and temperature.

Positive adsorption at the liquid surface implies that work must be done to transfer solute molecules from the surface to the bulk of the solution, which represents the free energy change due to the adsorption. If this free energy gain balances the free energy loss of the surface due to a change in surface tension or surface area, then the surface, and the foam of which it is part, is thermodynamically stable. The 'surface activity' of a solute can now be defined as the ability to lower the surface tension of a solution by transfer of solute molecules from the bulk solution to the surface.

3.5.1.2 Foam Drainage. It has been suggested (Ross, 1969) that protein solutions have surface layers that are 'plastic', meaning that they remain rigid under stress until the stress exceeds a yield value. Insofar as this is true, the bubble walls in a column of protein foam containing a fairly high ratio liquid to gas—known as a 'wet' foam—can be considered as effectively solid. Liquid then flows down through the interstices between the bubble walls with a velocity gradient in the liquid across the direction of flow such that the liquid velocity at the walls is zero. The rate of liquid drainage can then be calculated by assuming it to be the same as for liquid flow between two parallel plates (Ross, 1943), but even allowing for a change in the distance separating the plates with time, the resultant expression does not correlate well with experimental results.

The expression was modified in later work (Jacobi, Woodcock and Grove, 1956) to allow for the change in plate separation with time and with height in the foam column by successive reorientations of time and height co-ordinates. This enabled it to be used to describe a static, draining foam column where the foam wetness changes with height in the foam column. The expression obtained was of the form

$$\frac{dL}{dt} = b(a \cdot t + 1)^{-\frac{3}{2}} \tag{3.9}$$

where L is the volume of liquid draining from the column in time t, and a and b are constants. This expression fitted well to experimental results, and a very similar expression has been derived which gives a good correlation with other workers' results (Ross, 1969):

$$\frac{dL}{dt} = \tfrac{1}{2} \cdot B \cdot L_0 (B \cdot t + 1)^{-\frac{3}{2}} \tag{3.10}$$

where L_0 is the volume of liquid in the column at zero time and B is a constant.

The draining liquid can also be considered as flowing through a series of capillaries (Miles, 1945), which leads to an expression

$$(L_0 - L) - a \cdot \log (L_0 - L) + b = c \cdot t \tag{3.11}$$

where a, b and c are constants; this expression fitted well with experimental results.

3.5.1.3 Dynamic foam columns. The capillary drainage model has been extended to describe a dynamic foam in steady state (Haas and Johnson, 1967), where there is continuous upward vertical flow of foam counter-current to downward liquid drainage. A simple mass balance was carried out, but with the assumption that there was no gradient of foam wetness up the foam column. An expression for the average velocity, u, of liquid flowing in the capillaries was obtained from the Poiseuille equation for liquid flow in a pipe, corrected for the effective height of capillaries based on the volume fraction, ε, of liquid in the foam:

$$u = \frac{\rho g(1-\varepsilon)\delta^2}{32\mu} \tag{3.12}$$

where δ is the foam wall thickness, g is the gravitational acceleration factor, and ρ and μ are the density and viscosity of the liquid respectively. The fractional volume of gas in the foam is $(1-\varepsilon)$, so if the superficial gas velocity is G_s, then the upward velocity of bubbles will be $G_s/(1-\varepsilon)$. The number of capillaries in a cross-section is proportional to ε, so the nett area for liquid flow downwards, ε_c, is given by

$$\varepsilon_c = n(1-\varepsilon)\frac{\pi\delta^2}{4} \tag{3.13}$$

where n is the number of capillaries per unit cross-sectional area of the foam. From this, the nett liquid flow, per unit time, per unit cross-sectional area of the foam, L, is

$$L = u\cdot\varepsilon_c\,(\text{downwards}) - \frac{G_s\cdot\varepsilon}{(1-\varepsilon)}\,(\text{upwards}) \tag{3.14}$$

Since both ε and ε_c are volume fractions of liquid per unit volume of foam, then it can be assumed that there is a direct proportionality relation between them and

$$\varepsilon = k\cdot\varepsilon_c \tag{3.15}$$

where k is a constant independent of G_s, L or the bubble diameter, d. Combining equations (3.12), (3.13), (3.14) and (3.15),

$$L + \frac{G_s\cdot\varepsilon}{(1-\varepsilon)} = \frac{\rho g\varepsilon^2}{8nk^2\pi\mu} \tag{3.16}$$

By making a number of simplifying assumptions, such as the liquid density and viscosity being the same as for water and giving an arbitrary value of 1·5 to the constant k, an expression for foam density, ε_f, in terms of superficial gas velocity and bubble diameter, d, only was derived from equation (3.16):

$$\varepsilon_f = 3{\cdot}2\times10^{-4}\times\frac{G_s}{d^2} \tag{3.17}$$

With such simplifications, this sort of approach can apply only to a limited range of drainage conditions, and would not be valid for very short or very long draining times. With very long draining times, bubble walls become very thin and surface viscosity effects predominate over hydrodynamic effects.

3.5.1.4 More rigorous models. The models considered so far have been simplified by assuming that foam drainage occurs by flow through channels of circular cross-section or between parallel plates. The resultant expressions then contained empirical constants and were applicable only to the drainage of the wet, slow draining foams like those used in fire-fighting.

It has long been acknowledged that there are two distinct regimes of foam structure (Ross, 1967). Wet foams contain spherical, thick-walled bubbles that drain relatively quickly. As drainage progresses, the bubble walls become very thin, and pressure differences other than those resulting from hydrodynamic effects become significant, such as gradients in surface tension. As a result, the bubbles in the foam become polyhedral in shape (Plateau, 1873), being regular dodecahedra with faces bounded by narrow capillaries in a random alignment (Leonard and Lemlich, 1965). The bubble surfaces become very sharply curved at these boundaries, which are known as 'Plateau borders', and in this type of foam, most of the liquid content of the foam is held at the Plateau borders (de Vries, 1972). The true cross-sectional geometry of the capillaries formed by the Plateau borders and their orientation have been taken into account in an approach to foam drainage otherwise similar to that of Haas and Johnson discussed in Section 3.1.3 (Leonard and Lemlich, 1965). An expression for the effective number of capillaries per unit cross-sectional area of a foam was obtained by calculating the probability of a capillary being intersected by a horizontal plane and integrating over all possible angles of orientation of capillaries. The probable number of capillaries in a horizontal cross-section was $P_c/2$ per unit cross-sectional area, where P_c is the total length of capillaries per unit length of vertical foam column. From geometric considerations,

$$P_c = 7{\cdot}81(1-\varepsilon)/d^2 \tag{3.18}$$

so that the probable number of capillaries per unit cross-sectional area of the foam is $3{\cdot}9(1-\varepsilon)/d^2$. The interstitial liquid flow was then calculated by numerical integration of the general differential equation for momentum conservation, for rectilinear Newtonian flow of liquid through a typical capillary; the walls were taken as mobile and the liquid subject to surface viscosity. Integration of the local velocities thus calculated yielded a complex expression for the nett rate of liquid upflow, which is the foamate production rate. Satisfactory agreement with the foamate production rate from dry surfactant foams was obtained with the expression. However, as has been mentioned earlier, workers in the same school (Shih and Lemlich, 1971) found a considerable difference between the behaviour of detergent foams and those of biological surface-active materials, such as saponins and albumens.

3.5.2 Models for Solute Enrichment

Before considering models of foam fractionation systems, some attention will be given to two concepts fundamental to foaming processes, the surface excess and the nature of foam.

3.5.2.1 Surface adsorption. The fundamental phenomenon in foam fractionation is the formation of the surface excess, whose magnitude at equilibrium is defined by the Gibbs adsorption equation (Gibbs, 1928):

$$d\gamma = -R\cdot T\sum_i \Gamma_i\cdot d(ln\ a_i) \tag{3.19}$$

where Γ_i is the surface excess and a_i the activity of component 'i'. For a two-component solution, consisting of a solvent and only one solute, the expression for surface excess is then

$$\Gamma = -\frac{1}{R\cdot T}\cdot\frac{d\gamma}{d(ln\ C)} \tag{3.20}$$

where the activity of the solute is approximated to its concentration, C. The surface excess is measured in mass units per unit area, for example, g-mole/m^2. The change in surface excess with solute concentration is illustrated in Figure 3.7: surface excess rises with increasing concentration to a limiting constant value. At very low concentrations, the surface excess can be taken as being directly proportional to solute concentration; at the limiting value of surface excess, a saturated monolayer can be considered as formed. In practice, the usefulness of the Gibbs equation is somewhat limited: with a static surface, the time taken to reach equilibrium can be very long (Evans *et al.*, 1970), and in foaming processes where the surfaces move through the solution, hydrodynamic forces may induce uneven formation of the surface excess.

By assuming that the surface excess consists of a complete monolayer, its concentration can be estimated from the size and packing arrangement of the solute molecules (Lemlich, 1972). For example, the area of surface needed for the adsorption of a single BSA molecule has been reported by several workers (James and Augenstein, 1966): from accepted values of the molecular weight of BSA, the surface excess of BSA falls in the range 0·0011 to 0·0035 g/m^2.

3.5.2.2 Ideal foam. In considering models of surface enrichment, considerable simplification can be achieved by working in terms of an 'ideal foam'. Ideal foam is defined as foam in which

(a) all the bubbles are spherical;
(b) the liquid from which the foam originates is completely mixed;
(c) the liquid entrained in the foam, between the bubbles, has the same composition as the bulk liquid from which the foam originated;
(d) the bubbles have walls of composition of the surface excess, as calculated from the Gibbs equation, for a surface in equilibrium with the interstitial liquid of composition specified by assumption (c);
(e) the solutions are dilute;
(f) a single equilibrium stage is achieved (this will be explained in detail later).

A column of ideal foam can be considered as a stack of thin-walled hollow spheres, with the interstices between them containing the solution from which they were formed; the composition of the sphere walls is that of the surface excess in equilibrium with the interstitial liquid. When an ideal foam collapses, the resultant liquid has a volume taken to be that of the interstitial liquid, as the volume of the walls is negligible; the concentration of the foamate is due to both the surface excess layers and the interstitial liquid.

3.5.2.3 Continuous fractionation of a single solute. A model for the continuous fractionation in the simple mode (Figure 3.3) of a solution of a single surface-active species has been derived and tested experimentally using simple mass-balance equations (Newson, 1966). For a feed to the process of concentration C_0 and volume feed-rate V_0, foamate concentration C_f and volume production rate V_f, and residual product or 'raffinate' concentration C_r and volume production rate V_r, the mass-balance equations are:

$$V_0 = V_r + V_f \tag{3.21}$$

$$V_0 \cdot C_0 = V_r \cdot C_r + V_f \cdot C_f \tag{3.22}$$

$$V_f \cdot C_f = V_f \cdot C_r + A_s \cdot G \cdot \Gamma_r \tag{3.23}$$

where A_s is the specific surface of the foam (surface area per unit volume of foam) and Γ_r is the surface excess in equilibrium with the bulk solution entrained in the foam. Equations (3.21) and (3.22) express the conservation of liquid volume and solute mass, and equation (3.23) follows the definition of ideal foam. These equations can be combined to give the solute enrichment, E:

$$E = C_f/C_0 = 1 + \frac{G \cdot A_s \cdot \Gamma_r(V_0 - V_f)}{C_0 \cdot V_0 \cdot V_f} \tag{3.24}$$

Experiments with different solution feed-rates and gas flow-rates, using sodium dodecylbenzene sulphonate (NaDBS) as the surface-active solute, gave results in good agreement with predictions from the above equation: the values of Γ, the surface excess calculated from equation (3.24) were virtually the same as values calculated from surface-tension measurements and the Gibbs equation. The relation derived above is therefore valid for the simple mode insofar as the Gibbs equation is, so the degree of surfactant separation achieved was very close to that expected of a system in which equilibrium was reached between the solution and the surface excess. This is termed separation equivalent to a single equilibrium stage, as included in the definition of ideal foam. By making suitable assumptions, an expression for a 'stripper' arrangement was also derived, but without experimental verification.

A similar model was proposed by Lemlich (1968), but with the specific surface of the foam expressed in terms of bubble diameter. For spherical bubbles, the surface area in the foam per unit volume of gas (as opposed to foam) is $6/d$: where the simple foam fractionation system has been used to obtain values for the surface excess, then

$$\Gamma_r = \frac{(C_f - C_r) \cdot V_f \cdot d}{6G} \tag{3.25}$$

The values obtained for the surface excess from this kind of experimental system have been found to depend on the time allowed for drainage of the interstitial liquid in the foam.

3.5.2.4 Treatment of an ionic surfactant as two species. The model for fractionation of a single solute has been extended to consider the continuous removal of a cationic surfactant in terms of the two ions as separate species (Wood and Tran, 1966). The surfactant involved was ethylhexadecyldimethylammonium bromide (EHDABr), considered as $EHDA^+$ and Br^-. From the Gibbs equation (3.19),

$$d\gamma = -R \cdot T(\Gamma_{EHDA+} \cdot d \ln a_{EHDA+} + \Gamma_{Br-} \cdot d \ln a_{Br-}) \tag{3.26}$$

where Γ and a refer to the surface excesses and activities respectively of the species subscripted. For dilute solutions, and to preserve electrical neutrality,

$$a_{EHDA+} = a_{Br-} \simeq C_{EHDABr} \tag{3.27}$$

where C_{EHDABr} is the total concentration of surfactant; thus

$$\Gamma_{EHDA+} = \Gamma_{Br-} = \Gamma_{EHDABr} \tag{3.28}$$

and

$$\Gamma_{EHDABr} = \frac{1}{2RT} \cdot \frac{d\gamma}{d \ln C_{EHDABr}} \tag{3.28}$$

Following the same mass-balance procedure as in Section 3.2.3,

$$C_f - C_r = -\frac{3}{RTd} \cdot \frac{G}{V_f} \cdot \frac{d\gamma}{d \ln C_r} \tag{3.29}$$

$$C_0 - C_r = -\frac{3}{RTd} \cdot \frac{G}{V_0} \cdot \frac{d\gamma}{d \ln C_r} \tag{3.30}$$

$$C_f - C_0 = -\frac{3}{RTd} \cdot \frac{G\,V_r}{V_f \cdot V_0} \cdot \frac{d\gamma}{d \ln C_r} \tag{3.31}$$

and for a given gas flow-rate,

$$V_f(C_f - C_r) = V_0(C_0 - C_r) = \frac{V_0}{C_r}(C_f - C_0) \tag{3.32}$$

Qualified success was obtained in correlating these expressions with experimental data, and the limitations of the ideal foam model were inferred from the discrepancies. The value of the surface excess predicted by the above expressions was distinctly less than experimental values determined by other means for the wet foams obtained with high gas flow-rates. The inference was that the concentration of surfactant in the interstitial entrained liquid was lower than that of the bulk liquid. This effect is possible if the surface excess is incomplete as bubbles accumulate to form the foam: completion of the enriched layer could then occur at the expense of the entrained liquid, thus reducing its concentration. It is thought unlikely that such an effect would occur with macromolecular surfactants (Kanner and Glass, 1969). With low gas flow-rates, high values of surface excess were obtained: this was explained, more convincingly, as the effects of the coalescence of low-stability foam. The liquid released from the foam coalescence would flow back down the foam column, acting as an internal reflux stream, which would be expected to enhance enrichment beyond that predicted for a single equilibrium stage.

3.5.2.5 Multi-solute models. Separation of non-surface-active ions has been achieved by formation of complexes with an ionic surfactant as collector/foaming agent (Section 3.4). This technique has been particularly useful for the removal of metal ions, but most of the work reported has been of feasibility studies without a theoretical basis. However the ideal foam model, in the form applicable to polyhedral foams (Rubin, 1963), has been used to determine the separation factor, Γ/C, from the enrichment of metal-nitro complexes of mercury^{II} using EHDABr as collector (Miller and Sullivan, 1971). Foam fractionation with total recycle and with a constant foam column height was carried out with different concentrations of nitrate ions in the bulk solution. The separation factor was given by

$$\frac{\Gamma}{C} = (E-1)\cdot\frac{L\cdot d}{6{\cdot}59\,G} \tag{3.33}$$

and was reported as being analogous to the volume distribution factor in an ion-exchange column. The gas-liquid interface was considered as a mobile, two-dimensional ion-exchange medium where exchange could take place between, in this case, the nitrate ions and the metal nitro-complex, the positive charge centres resulting from the cationic surfactant adsorbed at the gas-liquid interface. The theoretical model was worked out using the appropriate equations and constants for ionic equilibrium (Karger and Miller, 1969), and the expression for the separation factor, Γ/C was found to agree qualitatively with values obtained for metal chloro-complexes in solutions rich in chloride ions, using EHDABr as the cationic collector. In the same way, EHDABr was considered as a collector for removal of nitrate and iodide ions by foaming their insoluble complexes: experiments with different foam column heights showed that no further adsorption or exchange occurred as the liquid drained from the foam.

No theoretical approach has apparently been developed specifically applicable to the foam separation of particulate material.

3.5.3 Discussion of Theoretical Considerations

Experimental verification of the theoretical treatments outlined above has been carried out with surfactants of the synthetic detergent type. The efficacy of separation of a foam fractionation process depends in the first instance on the behaviour of the surface-active solute at the gas-liquid interface, and it has been shown that synthetic polymers behave very differently at surfaces from ionic surfactants (Kanner and Glass, 1969). It may therefore be concluded that biological macromolecules will also show different surface behaviour from ionic surfactants.

The further enrichment occurring in the foam depends on the regime of foam structure, which in turn is determined by the operating conditions in the foaming process. In general, wet foam with spherical bubbles is formed with high superficial gas velocities and at low levels in the foam column,

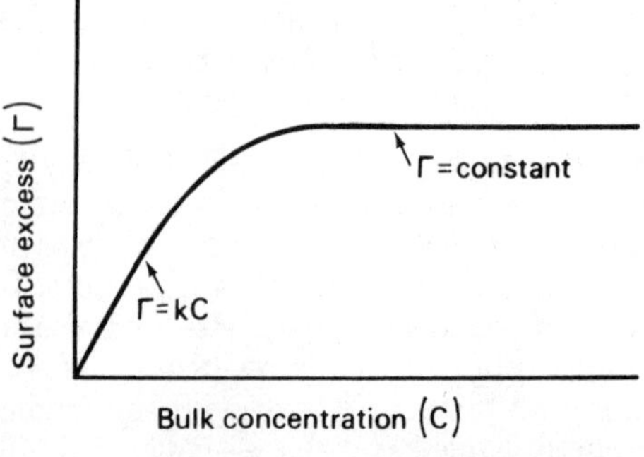

FIG. 3.7 Surface excess *v.* Bulk concentration.

even with inherently fast-draining foams; dry polyhedral foam results from low gas velocities and in high foam columns.

Increased enrichment is obtained in operating modes involving reflux of foamate; the appropriate equilibrium conditions for surface excess formation are presumably those in the region where the surface excess is proportional to solution concentration (Figure 3.7), with dilute solutions where the surface monolayer is incomplete (Lemlich, 1968). In an enriching or stripping column (Figures 3.4, 3.5, 3.6), there will be solute transfer between the foam surfaces and the downflowing liquid occurring continuously. The validity of this picture has, however, been challenged (Goldberg and Rubin, 1972); insofar as the model is valid, then the concept of equating a certain height of foam column to a single equilibrium stage is relevant—although this has also been questioned (Jashnani, 1973)—when the enrichment achieved is a function of the height of the foam column.

The alternative concept (Goldberg and Rubin, 1972) is that solute transfer in the enriching or stripping modes is concentrated at the base of the foam column adjacent to the pool of bulk solution and at the feed entry position (Figures 3.4, 3.5, 3.6), so that the length of the intermediary column has no effect on the separation efficacy of the system.

The equilibrium stage concept applies to very dilute solutions which are associated with the formation of polyhedral foams. In that regime, the bubble walls are under considerable tension (de Vries, 1972), in response to which there must be a transfer of solute (the so-called 'Marangoni effect'). The complexity of this regime is such that no model has been developed to take into account the interfacial forces, the solute transfer balancing these forces and the resultant shear stresses across the interface between the bubble wall and the interstitial liquid. These interfacial shear stresses define the parameter known as 'surface viscosity' which is a controlling factor in the flow of the interstitial liquid.

Even an apparently simple event like the coalescence of two bubbles has shown considerable complexity in taking into account the fluid flow and solute transfer involved in film thinning and eventual coalescence (Lee and Hodgson, 1968; Thomas, 1969). In assessing the applicability of the above concepts to the separation of biological surface-active materials, it is significant that even the most realistic model of foam drainage so far developed (Leonard and Lemlich, 1965) could be fitted to the behaviour of foams of saponin and BSA only when quite untypical values of the surface viscosity were used (Shih and Lemlich, 1967).

When the concentration of surface-active material in the solution is sufficiently high to ensure that the interfacial monolayers are complete, the equilibrium conditions discussed above become irrelevant. If the bubble geometry remains constant, so that the specific surface of the foam is unchanged, then the concentration of solute in the foamate will depend largely on the volume of entrained solution diluting the constant amount of solute in the interfacial monolayers, as in the situation described by Grieves (1970).

3.6 FOAM FRACTIONATION OF PROTEINS

3.6.1 Properties of Protein Foams

Protein solutions in general are well-known for their property of forming very stable foams, and are widely used industrially as foam stabilizers. This stability is attributed to the rigidity of the interfacial layers (Kanner and Glass, 1969), from which it may be inferred that the monolayers are saturated with protein. The precise molecular orientation is the subject of some controversy (James and Augenstein, 1966). Except under limiting conditions of high foam columns and low gas flow-rate, protein foams tend to be wet and consist of spherical, thick-walled bubbles. This is very much in accordance with the definition of ideal foam, and enrichment in the foam can be expected to occur by drainage of interstitial liquid from between the bubbles, rather than according to the complex models developed to describe the behaviour of polyhedral foams.

3.6.2 A Simplistic Model for Foam Fractionation of Protein

Following the above precepts, the authors have developed a simplistic model for foam fractionation which has correlated well, on a semi-empirical basis, with the results from the foam fractionation of BSA solutions in a simple batch system (Thomas and Winkler, unpublished work). The mechanism of enrichment is that assumed for ideal foam: formation of complete surface excess layers on bubble interfaces, entrainment of bulk solution into foam with the bubbles followed by increasing enrichment with progressive drainage of the entrained liquid from the foam. Empirical expressions for the volume of liquid entrained into the foam and for the subsequent rate of drainage out of the foam were then developed, the values of the empirical constants being found by curve-fitting methods.

Suppose a single bubble of diameter d has a surface excess Γ and carries into the foam a volume V_b of bulk solution of concentration C: the foamate from a number of such bubbles will then have a concentration C_f, given by

$$C_f = \frac{\pi d^2 \Gamma + V_b \cdot C}{V_b} \tag{3.34}$$

assuming that the volume of the surface excess layer is negligible compared with the volume of entrained solution. The resulting enrichment ratio, E, is then

$$E = \frac{C_f}{C} = 1 + \frac{\pi d^2 \Gamma}{V_b \cdot C} \tag{3.35}$$

Measurements of the volume production rate of foamate indicated an approximately linear relationship between this quantity and the bulk liquid protein concentration over a range of 50 to 600 g BSA/m^3. Equation (3.35) was then modified to

$$E = 1 + \frac{\pi d^2 \Gamma}{C(a + b \cdot C)} \tag{3.36}$$

where a and b are constants depending on the operating conditions.

The values of a and b were determined, for a particular set of operating conditions, by linear regression analysis of the enrichment (E)—concentration (C) data for batch runs; the bubble diameter was measured photographically and a value of surface excess, calculated as described in Section 3.5.2.1, of 0·0024 g/m^2 was assumed.

Further enrichment results from the drainage of entrained liquid: the more drainage that occurs, the less are the surface excess layers diluted in the foamate. This mechanism is assumed to continue until the foam is dry enough to form polyhedral foam, and the ideal foam model is no longer valid. The amount of entrained liquid draining from the foam depends on the drainage rate and the time for which drainage occurs.

An individual foam element can be assumed to be draining throughout its life, from its formation through its progress up the foam column to the foam collection level. If the foam column height is H, the superficial gas velocity G_s and the volume fraction of gas in the foam F, then the age of a foam element at height H is t, given by

$$t = \frac{H}{G_s \cdot F} \tag{3.37}$$

The volume fraction of gas is given by

$$F = \frac{\pi d^3}{\pi d^3 + 6V_b} \tag{3.38}$$

but in practice, values of F calculated from this expression were usually about 0·95: in view of the other approximations made in this treatment, F can reasonably be taken as unity.

Values of V_b, the entrained liquid volume, calculated from equation (3.35) using data at different foam column heights, decreased with increasing foam height, but at a diminishing rate. Considering foam as a stack of spheres from between which the interstitial liquid is draining, then the closer the spheres are to each other, the greater will be the resistance to the flow of the interstitial liquid.

Since the separation of the spheres is determined by the proportion, by volume, of interstitial liquid in the foam, then the drainage rate (the rate of removal of interstitial liquid) is proportional to the volume of liquid present at a given time. This implies that the volume of interstitial liquid decays exponentially with time: a similar conclusion has been reached by other workers (de Vries, 1972). If the volume of interstitial liquid at foam age t is V_t, then an exponential decay expression will be of the form

$$V_t = V_l \cdot e^{-Z \cdot t} \tag{3.39}$$

if it is assumed that the entrained volume on formation is V_l; both V_t and V_l refer to the entrained volumes associated with a single bubble. The parameter Z is likely to be dependent on protein concentration as a result of its effect on the viscosity of the interstitial liquid. The composition of the interstitial liquid at any point in the foam is not easy to calculate, as it is a mixture of the original entrained liquid with that draining down on to it; however, it must bear some relation to the bulk liquid protein composition.

Combining equations (3.36), (3.37) and (3.39), and taking the volume fraction of gas as being close to unity, then

$$E = 1+\frac{\pi d^2 \Gamma \cdot e^{Z \cdot H/G_s}}{C(a+b \cdot C)} \tag{3.40}$$

Where concentration-enrichment data were available for different foam column heights, but at the same superficial gas velocity, covering the same range of bulk liquid concentrations, the empirical constants could be calculated from data at two different foam column heights. These values can then be used to predict enrichment-concentration data at any other foam column height, for the same operating conditions. The predictions gave good agreement with measured values (Figure 3.8).

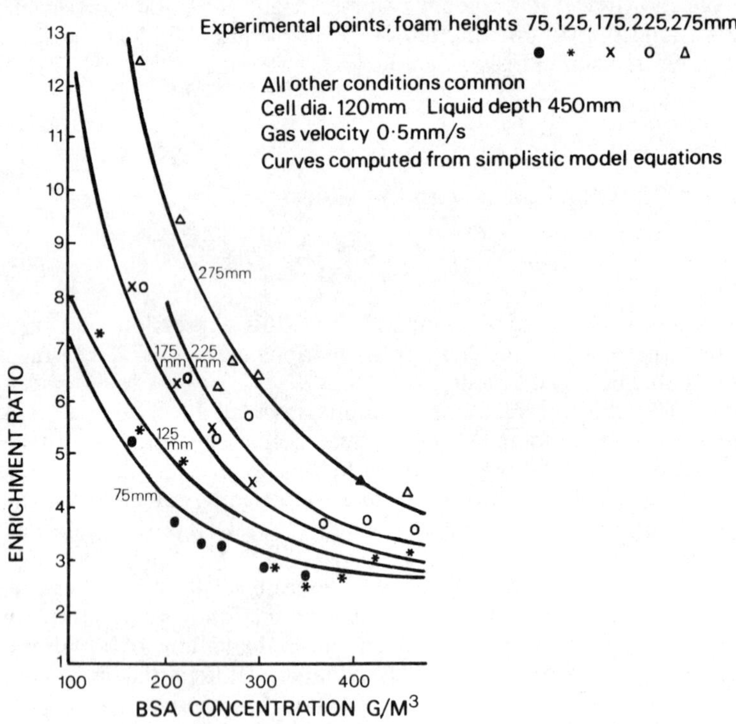

FIG. 3.8 Enrichment *v*. B.S.A. concentration, different foam column heights.

3.6.3 Limitations of the Simplistic Model

The original aim of the simplistic approach was to produce a design equation that describes the enrichment achieved in a foaming operation over a wide range of conditions. While equation (3.40) has been found to give accurate predictions of the performance of a foaming cell of a particular diameter working at a particular superficial gas velocity, the values of the empirical constants, *a*, *b* and *Z* have been found to vary with cell diameter. Correction factors to allow for these variations can be introduced, on an empirical basis, and have met with some success in correlating with other experimental results.

3.7 CONCLUSIONS

Foam fractionation has been shown by several different groups of workers to give useful concentration of enzymes and other proteins, separation of enzymes from each other and enzymes from adventitious proteins in their preparations; froth flotation has given successful concentration of microbial cultures and separation of spores from autolysed cultures. Operating conditions for these processes can be selected so that virtually no denaturation of protein occurs. The complexity of foaming systems has meant that a fully quantitative model of the process does not yet exist, but some success has been found with semi-empirical models.

REFERENCES

Andrews, G. and Schütz, F. (1945), *J. Biochem.*, **39**, 1.

Bader, R. and Schütz, F. (1946), Trans. Faraday Soc., **42**, 571–576.

Black, S. H., MacDonald, R. E. Hashimoto, T. and Gerhardt, P. (1960), *Nature*, **185**, 782–783.

Boyles, W. A. and Lincoln, R. E. (1958), *Appl. Microbiol.*, **6**, 327–334.

Bretz, H. W., Grieves, R. B. and Wang, S. L. (1966), *Appl. Microbiol.*, **14**(5), 778–783.

Buckingham, J. H. (1970), *J. Sci. Food Agr.*, **21**, 441–445.

Charm, S. E. (1972), in *Adsorptive Bubble Separation Techniques.* Ed. by R. Lemlich, AP, Ch. 9, 157–174.

Charm, S. E., Morningstar, J., Matteo, C. C. and Paltiel, B. (1966), *Anal. Biochem.*, **15**, 498–508.

Courrier, R. and Dognon, A. (1939), *C. R. Acad. Sci.* Paris, **202**, 242.

Derjauguin, B. V. and Dukhin, S. S. (1960), *Trans. I.M.M.*, **70**, 221–246.

de Vries, A. J. (1972), in *Adsorptive Bubble Separation Techniques.* Ed. by R. Lemlich, AP, Ch. 2, 7–30.

Dognon, A. (1941), *Rev. Sci.* (France), **79**, 613–619.

Dognon, A. and Gougerot, L. (1947), *Bull. Soc. Chim. Biol.*, **29**, 702.

Dorman, D. C. and Lemlich, R. (1965), *Nature*, **207**, 145.

Evans, M. T. A., Mitchell, J., Musselwhite, P. R. and Irons, L. (1970), in *Surface Chemistry of Biological Systems.* Ed. by M. Blank, Plenum Press (NY), 1.

Gaudin, A. M. (1957), *Flotation*, McGraw-Hill.

Gaudin, A. M., Davis, N. S. and Bangs, S. E. (1962a), *Biotech. Bioeng.*, **4**, 211–222; (1962b), ibid. 223.

Gaudin, A. M., Mular, A. L. and O'Connor, R. F. (1960a), *Appl. Microbiol.*, **8**, 84–90; (1960b), ibid. 91–97.

Gibbs, J. W. (1928), *Collected Works*, Longmans.

Goldberg, M. and Rubin, E. (1972), *Sep. Sci.*, **7**, 51–73.

Grieves, R. B. and Bhattacharya, D. (1970), *Sep. Sci.*, **5**(5), 583–601.

Grieves, R. B. and Wang, S. L. (1966), *Biotech. Bioeng.*, **8**, 323–336.

Grieves, R. B. and Wang, S. L. (1967a), *Appl. Microbiol.*, **15**(1), 76–81.

Grieves, R. B. and Wang, S. L. (1967b), *Biotech. Bioeng.*, **9**, 187–194.
Haas, P. A. and Johnson, H. F. (1967), *Ind. Eng. Chem. Fund.*, **6**, 225–233.
Harper, D. O. and Lemlich, R. (1966), *A.I.Ch.E.J.*, **12**, 1220.
Holmström, B. (1968), *Biotech. Bioeng.*, **10**, 551–552.
Hopper, S. H. and McCowen, M. C. (1952), *J. Am. Water Works Assn*, **44**, 719–726.
Iibuchi, S., Yano, T. and Torikata, Y. (1974), *Agr. Biol. Chem.*, **38**(2), 395
Jacobi, W. M., Woodcock, K. E. and Grove, C. S. (1956), *Ind. Eng. Chem.*, **48**, 2046–2051.
James, L. K. and Augenstein, L. G. (1966), *Adv. Enzymol.*, **28**, 1–40.
Jashnani, I. L. and Lemlich, R. (1973), *Ind. Eng. Chem. Proc.*, **12**, 312–321.
Jenkins, D., Scherfig, J. and Eckhoff, D. W. (1972), in *Adsorptive Bubble Separation Techniques.* Ed. by R. Lemlich, AP, Ch. 14, 219–242.
Jones, W. T. and Lyttleton, J. W. (1972), *N.Z. J. Agric. Res.*, **15**(3), 506.
Kanner, B. and Glass, J. E. (1969), *Ind. Eng. Chem.*, **61**(5), 31–41.
Karger, B. L. (1966), *US Naval Res. Contr. File AD*, 631–995.
Karger, B. L., Grieves, R. B., Lemlich, R., Rubin, A. J. and Sebba, F. (1967), *Sep. Sci.*, **2**, 401–404.
Karger, B. L. and Miller, M. W. (1969), *Anal. Chem. Acta*, **48**(2), 273–290.
Klassen, V. I. and Mokrusov, V. A. (1963), *An Introduction to the Theory of Flotation*, Butterworth.
Lauwers, . (1964), *Verh. Kon. Vlaam. Acad. Wetens. Belg.*, **78**(5), 26.
Lee, J. C. and Hodgson, T. D. (1968), *Chem. Eng. Sci.*, **23**, 1375–1397.
Lemlich, R. (1968), *Ind. Eng. Chem.*, **60**(10), 16–29.
Lemlich, R. (1972), *Adsorptive Bubble Separation Techniques.* Ed. by R. Lemlich, AP, Ch. 3, 33–51; Ch. 7, 133–143.
Leonard, R. A. and Lemlich, R. (1965), *A.I.Ch.E.J.*, **11**, 18–25.
London, M., Cohen, M. and Hudson, P. B. (1954), *Biochim. Biophys. Acta*, **13**, 111–120.
London, M. and Hudson, P. B. (1953), *Arch. Biochem. Biophys.*, **46**, 141.
Mahler, H. R. and Cordes, E. H. (1966), *Biological Chemistry*, Harper, 510.
Miles, G. D., Shedlovsky, L. and Ross, J. (1945), *J. Phys. Chem.*, **49**, 93–107.
Miller, M. W. and Sullivan, G. L. (1971), *Sep. Sci.*, **6**(4), 553–558.
Miyazu, Y. and Yano, T. (1974), *Agr. Biol. Chem.*, **38**(1), 183–188.
Moore, E. W. and Bryant, T. G. (1954), *US Army Res. Contr.*, DA49007-MD317.
Nakagaki, M. (1957), *J. Phys. Chem.*, **61**, 1266–1270.
Newson, I. H. (1966), *J. Appl. Chem.*, **16**, 43–49.
Ostwald, W. and Siehr, A. (1937), *Kolloid Z.*, **79**, 11.
Ostwald, W. and Mischke, M. (1940), ibid. **90**, 205.
Ouichi, K. and Akuyama, H. (1971), *Agr. Biol. Chem.*, **35**, 1024–1032.
Ouichi, K. and Numokawa, Y. (1973), *J. Ferment. Technol.*, **51**, 85.
Plateau, J. A. F. (1873), *Statique des Liquides*, Paris.
Rosen, G. D. (1971), *Poultry Ind. Apr.*, 21–22.
Ross, S. (1967), *Chem. Eng. Progr.*, **63**(9), 41–47.
Ross, S. (1969), *Ind. Eng. Chem.*, **61**(10), 48–57.
Rubin, A. J. (1968), *Biotech. Bioeng.*, **10**, 89–98.
Rubin, A. J. *et al.* (1966), ibid. **8**, 135–151.
Rubin, E. (1963), Ph.D. Dissertation, Columbia Univ.

Schnepf, R. W. and Gaden, E. L. (1959), *J. Biochem. Microbiol. Tech. Eng.*, **1**, 1–11.
Schütz, F. (1942), *Trans. Faraday Soc.*, **38**, 85–93.
Schütz, F., Bader, R. and Stacey, M. (1944), *Nature*, **154**, 183.
Shah, G. N. and Lemlich, R. (1970), *Ind. Eng. Chem. Fund*, **9**, 350.
Shih, F. S. and Lemlich, R. (1971), ibid. **10**, 254–259.
Stephan, D. G. (1965), *Civil Eng.*, **35**(9), 46–49.
Tait, P. G. (1885), *Properties of Matter*, Black, Edinburgh.
Thomas, A. (1969), M.Sc. Thesis, Univ. of Wales.
Wallace, G. T. and Wilson, D. F. (1969), *US Naval Res. Lab*, 6958 AD 697 924.
Wood, R. K. and Tran, T. (1966), *Can. J. Chem. Eng.*, **44**, 322–326.
Zwierkowski, W. and Medrzyka, K. B. (1973), *Sep. Sci.*, **8**, 57–69.

Chapter **4**

Aeration of Mould and Streptomycete Culture Fluids

G. T. BANKS, Department of Biochemistry, Imperial College, London

SYMBOLS AND NOMENCLATURE

A	= Area
a	= Gas/liquid interfacial area
C	= Dissolved oxygen concentration in bulk of liquid
C^*	= Dissolved oxygen concentration in equilibrium with partial pressure of oxygen in gas phase
c	= Constant dependent on vessel geometry
D	= Impeller diameter
D_L	= Diffusivity of oxygen in liquid phase
D_t	= Tank diameter
F	= Applied force
g	= Gravitational constant
H_v	= Vortex depth
K	= Consistency coefficient
K_c	= Casson viscosity
K_L	= Mass transfer coefficient
K_La	= Volumetric transfer coefficient (aeration efficiency)
k	= Proportionality factor of Metzner and Otto
k', k'' etc.	= Constants
L	= Length (height) of impeller blades
M	= Momentum factor
N	= Impeller rotational speed
N_a	= Aeration Number (dimensionless)
N_{De}	= Deborah Number (dimensionless)
N_{Fr}	= Froude Number (dimensionless)
N_P	= Power Number (dimensionless)
N_{Re}	= Reynolds Number (dimensionless)
N_{We}	= Weber Number (dimensionless)
n	= Flow behaviour index
OSR	= Oxygen solution rate
P	= Power consumption in un-gassed liquid
P_g	= Power consumption in gassed liquid
Q	= Volumetric air flow rate

r = Impeller radius
S = Impeller tip speed
T = Torque
t_m = Mixing time
V = Volume of liquid
V_s = Superficial (linear) air velocity
V_t = Terminal rising velocity of bubbles

γ = Shear rate
$\bar{\gamma}$ = Average shear rate
μ = Viscosity (liquid)
μ_a = Apparent viscosity
$\bar{\mu}_a$ = Average apparent viscosity
μ_g = Viscosity of gas phase
μ_L = Viscosity of suspending liquid
μ_s = Viscosity of suspension
ϕ = Volume fraction of suspended particles
ρ = Liquid density
σ = Surface tension
τ = Shear stress
$\bar{\tau}$ = Average shear stress
τ_0 = Yield stress
λ = Characteristic material time after Prest *et al.*
η = Coefficient of rigidity

4.1 INTRODUCTION

The normal method of aerating microbial cultures on the laboratory scale is by means of the shaken flask fermentation. The culture (usually 50–1000 mls) is contained within a suitable flask which is shaken mechanically with a reciprocating or rotary motion; oxygen transfer into the culture fluid takes place by a process of diffusion through the surface liquid film, which is being constantly renewed under these conditions. This method of culture aeration is adequate, both in terms of efficiency of oxygen transfer and reproducibility, for the majority of microbial cultures, including mould and streptomycete cultures.

As the scale of operation increases, however, this method of culture aeration becomes impracticable, since the liquid surface area/volume ratio decreases with increasing scale. Pilot and production scale submerged culture fermentations are conventionally carried out in baffled, stirred, sparger-aerated, cylindrical vessels of up to 500 cubic metres capacity; the design of such vessels has been described elsewhere (Holland, F. A. and Chapman, F. S., 1966). Although the technique of sparger aeration is very efficient, the difficulties in supplying adequate quantities of oxygen to large scale fermentations of this size are considerable and represent one of the major problems confronting fermentation technologists and biochemical engineers in the field of fermenter design, optimization of fermentation conditions and scale-up. It is not surprising, therefore, that the subject of culture aeration has attracted a great deal of attention over the last 40 years from both industrialists and academics; evidence of this interest lies in the considerable number of publications which have appeared in the scientific literature. However, most of the reported research has concerned the aeration of non-viscous Newtonian fluids, i.e. fluids which obey Newton's law of viscous flow; the culture fluids of most unicellular micro-organisms (e.g. bacteria and yeasts) fall into this category. It is probably true to say that aeration phenomena in such fluids are relatively well understood.

Unfortunately, from the point of view of the industrialist, the overwhelming majority of industrially important microbial metabolites are not produced by unicellular micro-organisms but by filamentous multi-cellular micro-organisms such as moulds and streptomycetes. The culture fluids of these organisms are both very viscous and non-Newtonian in behaviour. Until quite recently the aeration of such fluids was poorly understood, to such an extent that a fermentation technologist or biochemical engineer, faced with the problem of predicting the effects of operating variables on culture aeration or scaling-up, would probably have had to resort to trial-and-error methods rather than attempting to solve these problems on a rational basis. However, over the last 15–20 years, important developments have taken place which have increased our knowledge and understanding in this area. It is the purpose of this article to explain and review some of the more important of these developments.

4.2 AERATION-AGITATION OF NON-VISCOUS NEWTONIAN FLUIDS

Before discussing problems associated with the aeration of mould and streptomycete fluids, it is necessary very briefly to summarize the most significant relationships which have been established with non-viscous Newtonian fluids.

In submerged culture fermentations, oxygen (a sparingly soluble gas) must first enter solution before it can reach the microbial cells. The rate of solution is governed by the following equation, which is typical of any mass transfer process.

$$OSR = K_L a\,(C^* - C) \tag{4.1}$$

where OSR = Oxygen solution rate
K_L = Mass transfer coefficient
a = Gas/liquid interfacial area
C^* = Concentration of dissolved oxygen in equilibrium with partial pressure of oxygen in gaseous phase
C = Concentration of dissolved oxygen in bulk of liquid

Since it is extremely difficult to determine both the gas/liquid interfacial area and, in a sparger-aerated fermenter, the mass transfer coefficient independently, the two terms K_L and a are usually combined to form $K_L a$, the volumetric transfer coefficient. The volumetric transfer coefficient is almost universally employed as a measure of aeration efficiency, since it is independent of dissolved oxygen concentration; it has the dimension of reciprocal time and the normal units employed are hr^{-1}. $K_L a$ may be regarded as the reciprocal of the overall supply-side resistance to oxygen transfer. The oxygen concentration difference (expression in brackets in equation (4.1)) represents the driving force across this resistance. In an oxygen limited fermentation, the dissolved oxygen concentration in the bulk of the fluid is usually close to zero, in which case equation (4.1) reduces to:

$$OSR = K_L a\, C^* \tag{4.2}$$

To be exact, in an oxygen limited fermentation, it has been shown (Winzler, R. J., 1941) that the dissolved oxygen concentration falls below a critical value (C_{crit}), which is usually substantially below 10% of the saturation value.

Many factors are known to affect aeration efficiency ($K_L a$), including such parameters as agitation, air flow rate, air pressure, temperature, vessel geometry, fluid characteristics, presence of antifoam agents, etc. By far the most important of these is the degree of agitation. Agitation facilitates oxygen transfer in stirred fermenters in basically four ways (see Figure 4.1).

(i) By dispersing the air in the culture fluid in the form of small air bubbles, it increases the area available for oxygen transfer
(ii) It delays the escape of air bubbles from the fluid (i.e. it increases the bubble retention time and gas 'hold-up')

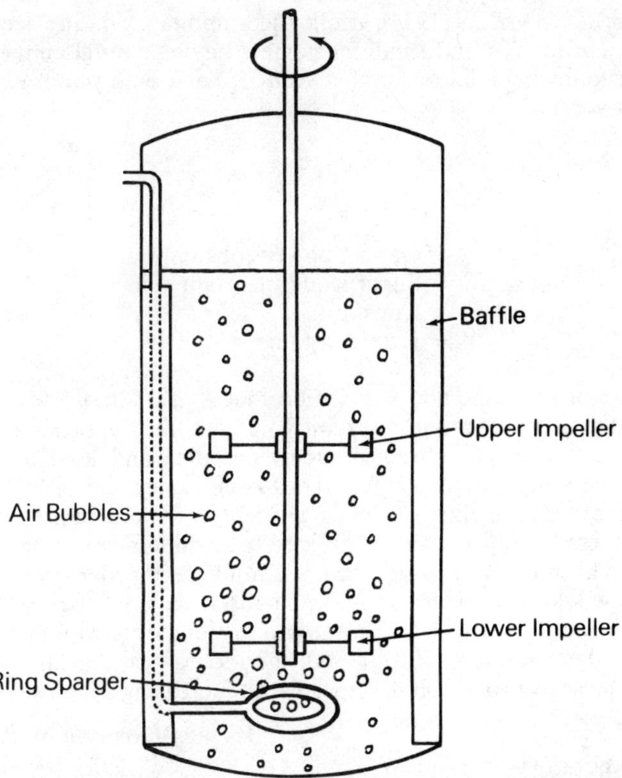

FIG. 4.1 Diagram illustrating technique of sparger aeration in vessel with multiple impellers.

(iii) It prevents the coalescence of air bubbles
(iv) By creating turbulence in the culture fluid, it decreases the thickness of the liquid film at the gas/liquid interface.

In a fermenter with two or more impellers, the bottom impeller is primarily concerned with promoting factors (i) and (iv) above, whilst the upper impellers are mainly concerned with promoting factors (ii), (iii) and (iv). All the available evidence strongly suggests that the primary resistance to oxygen transfer into solution lies in the liquid film at the gas/liquid interface and not in the gas film. Accordingly, fermenters are designed to promote turbulence in the liquid phase rather than the gaseous phase. In non-viscous Newtonian systems, it is equally unlikely that the liquid film at the cell surface is rate-limiting, except where 'clumping' or cell aggregation occurs (Phillips, D. H. and Johnson, M. J., 1961). The degree of agitation is usually represented by the power consumed in stirring the fluid or, less frequently, by the mixing time (t_m). Power consumption can be measured by a dynamometer attachment to the fermenter, by a strain gauge attached to the agitator shaft or, in the case of large fermenters, by measuring the electrical power consumption: the mixing time is determined by measuring the time taken to distribute some factor uniformly throughout the fermenter, e.g. a dye, acid or base, salt, heat, etc.

Cooper, C. M. *et al.* (1944), using the sulphite oxidation technique for measurement of K_La and small fermenters having a single impeller, established the following relationship between K_La and both power consumption and air flow rate:

$$K_La = k' \left(\frac{Pg}{V}\right)^{0\cdot95} V_s^{0\cdot67} \tag{4.3}$$

where
P_g = Gassed power consumption
V = Superficial (linear) air velocity
k' = Constant
V = Liquid volume

It will be seen from the above equation that K_La is almost directly proportional to the power consumption/unit volume. This proportionality between K_La and power consumption is widely accepted and used to predict the effect of operating variables upon aeration efficiency and in scale-up. However, this proportionality is almost certainly of more limited application than is generally realized, being confined to small fermenters having a single impeller. This is not surprising since, in a multiple impeller system, the upper impellers would be expected to make a substantial contribution to the total power consumption without making a proportionate contribution to aeration efficiency. Bartholomew, W. H. (1960) showed, in fact, that the exponent on Pg/V in the above equation decreased with increasing scale as follows:

Scale	*Value of Exponent on Pg/V*
Laboratory	0·95
Pilot Plant	0·67
Production Plant	0·50

The lower value of the exponent on Pg/V in fermenters with two or more impellers has subsequently been confirmed by various workers (e.g. Taguchi, H. and Kimura, T., 1970). It should be noted that the term V_s (superficial air velocity) is used in equation (4.3) above rather than the volumetric air flow rate. The former term is preferred since it is related to gas hold-up. The superficial air velocity can be calculated by dividing the volumetric air flow rate by the cross-sectional area of the vessel; the usual units for V_s are ft/sec. or cm/sec. Other workers have found a similar relationship between the volumetric transfer coefficient and superficial air velocity. It should be noted that the experimental work of Cooper *et al.*, was carried out in geometrically similar vessels in the region of fully developed turbulent flow and the established relationships only apply under these conditions.

Rushton, J. H. *et al.* (1950), by applying the technique of dimensional analysis, derived the following expression, relating to conventionally designed stirred reactors:

$$N_P = c(N_{Re})^x (N_{Fr})^y \tag{4.4}$$

or

$$\frac{P}{\rho N^3 D^5} = c\left(\frac{\rho N D^2}{\mu}\right)^x \left(\frac{N^2 D}{g}\right)^y \tag{4.5}$$

where $N_P = P/\rho N^3 D^5$ = Power Number
$N_{Re} = \rho N D^2/\mu$ = Reynold's Number
$N_{Fr} = N^2 D/g$ = Froude Number
c = Constant dependent on vessel geometry but not size
ρ = Liquid density
μ = Liquid viscosity
g = Gravitational constant
N = Impeller rotational speed
D = Impeller diameter
P = Power consumption for un-gassed liquid

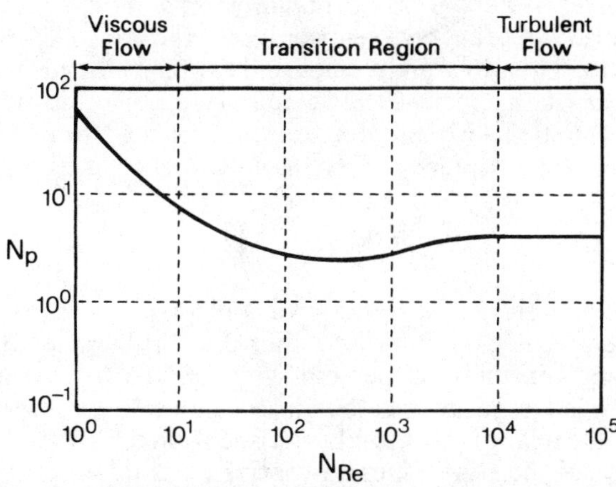

FIG. 4.2 A typical power curve.

The Power Number, Reynold's Number and Froude Number are all dimensionless groups or numbers used in chemical engineering calculations, representing the ratio of applied to opposing forces within the system. In a fully baffled vessel, gravitational forces exert a negligible effect and the Froude Number can thus be ignored; equation (4.4) therefore reduces to:

$$N_P = c(N_{Re})^x \tag{4.6}$$

A plot of Power Number against Reynold's Number on log-log co-ordinates is known as a power curve and may be constructed from experimentally obtained data. A typical power curve, similar to those obtained by Rushton *et al.* is shown in Figure 4.2. It should be noted that a given power curve applies to vessels of identical geometry, irrespective of size. It will be seen from Figure 4.2 that a typical power curve can be divided into three clearly defined regions, as shown below:

Region	*Reynold's Number*	*Slope of Plot* (x)
Viscous Flow	$< 10^1$	-1
Transition	$10^1 - 10^4$	Variable
Turbulent Flow	$> 10^4$	0

The Reynold's Number is thus an indicator of the degree of turbulence within the system and may be used to reveal the flow regime which exists. By substituting the experimentally determined values of the exponent x relating to the regions of viscous and turbulent flow, into equation (4.6), the two following equations can be simply derived:

$$P = c\rho N^3 D^5 \qquad \text{(Turbulent flow)} \tag{4.7}$$

$$P = c\mu N^2 D^3 \qquad \text{(Viscous flow)} \tag{4.8}$$

These equations are widely used to predict the effects of operating variables upon power consumption in stirred reactors and in scale-up. The former equation is more relevant to fermentation, since conditions of viscous flow seldom exist in fermenters. If proportionality (or, at least, a known relationship) is assumed to exist between aeration efficiency ($K_L a$) and power consumption, this equation can be employed to predict the effect of operating variables on aeration efficiency. The following equation, which can be derived from equation (4.7) can, for example, be used to scale-up in geometrically similar fermenters on the basis of constant power input/unit volume:

$$N_2 = N_1 \left(\frac{V_1}{V_2}\right)^{\frac{2}{9}} \tag{4.9}$$

where the suffices 1 and 2 refer to the small and large fermenters respectively. The rotational speed of the impeller required in the large fermenter to give the same power consumption per unit volume (and hence same value of $K_L a$) as in the small fermenter can thus be calculated. In the transition region, no similar simple relationships can be derived from equations (4.5) and (4.6) since the slope of the power curve is variable; in this region, a complete power curve must be constructed and utilized for prediction and scale-up studies.

The work of Rushton *et al.* relates only to power consumption in un-gassed fluids. Gassing (i.e. aeration) reduces power consumption appreciably, since the presence of air bubbles, particularly in the region of the impellers, reduces the density of the liquid/gas mixture. Aeration can reduce power consumption to as little as one-third of its original value (see Figure 4.3). Oyama Y. and Endoh, K. (1955) investigated the effect of gassing on power consumption. They proposed that the ratio of gassed to un-gassed power consumption could be related to another dimensionless group, the Aeration Number (N_a)

i.e.

$$\frac{P_g}{P} = f(N_a)$$

where

$$N_a = \frac{V_s}{ND} = \frac{\text{Superficial air velocity}}{\text{Impeller tip speed}}$$

A plot of the ratio of gassed to un-gassed power against Aeration Number, according to Oyama and Endoh, is given in Figure 4.3. Michel, B. J. and

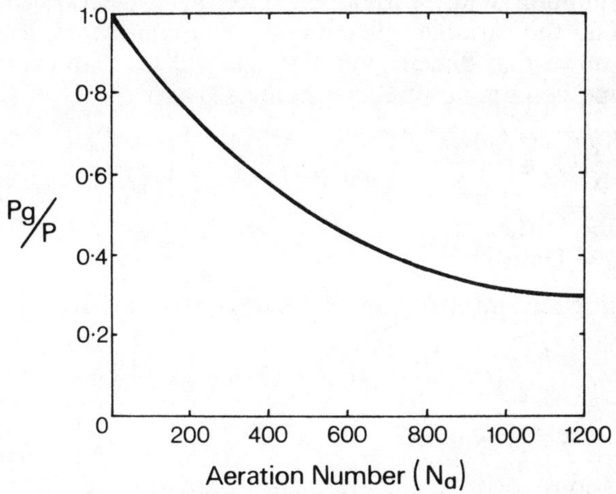

FIG. 4.3 The effect of gassing upon power consumption. [*Oyamah, Y. and Endoh, K., 1955.*

Miller, S. A. (1962) established a better and more comprehensive, although empirical, correlation for gassed power consumption, as indicated below:

$$Pg = k'\left(\frac{P^2ND^3}{Q^{0\cdot56}}\right)^{0\cdot45} \tag{4.10}$$

where Q = Volumetric air flow rate.

A graph, illustrating the final correlation of Michel and Miller, which applies to the turbulent flow region, is depicted in Figure 4.4. The group in parentheses in equation (4.10) is not dimensionless.

The afore-mentioned well established relationships between aeration efficiency (K_La), power consumption and operating variables are not the only correlations for K_La. Indeed, they cannot be comprehensive since they

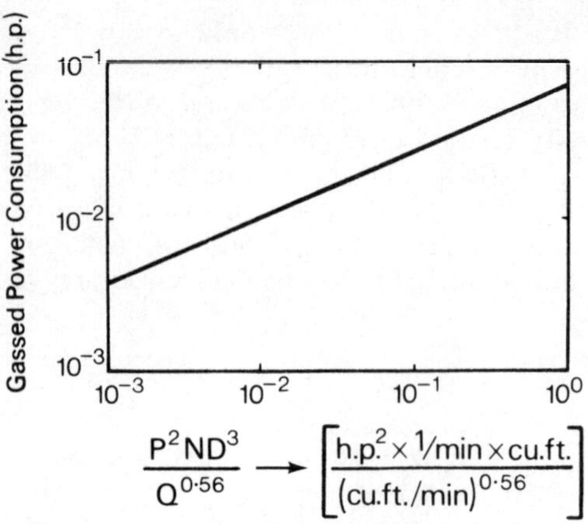

FIG. 4.4 Correlation for gassed power consumption. [*Michel, B. J. and Miller, S. A., 1962.*

exclude certain liquid properties, such as surface tension and diffusivity, which are known to affect oxygen transfer. Also, it is not necessary to include power consumption in any correlation since this is determined by the other variables. If all the variables affecting $K_L a$ are considered, Richards, J. W. (1961) has shown that dimensional analysis will yield an expression of the following type, containing six dimensionless groups:

$$\frac{K_L a}{N} = k'\left(\frac{\rho N D^2}{\mu}\right)^{\alpha}\left(\frac{\sigma}{\rho N^2 D^3}\right)^{\beta}\left(\frac{V_s}{ND}\right)^{\gamma}\left(\frac{D_L}{ND^2}\right)^{\delta}\left(\frac{N^2 D}{g}\right)^{\varepsilon} \tag{4.11}$$

or where σ = Surface tension
D_L = Diffusivity

or, substituting recognizable named dimensionless numbers

$$\frac{K_L a}{N} = k'(N_{Re})^{\alpha}(N_{We})^{\beta}(N_a)^{\gamma}\left(\frac{D_L}{ND^2}\right)^{\delta}(N_{Fr})^{\varepsilon} \tag{4.12}$$

where N_{We} = Weber Number

Such expressions serve to illustrate the complexity of the problem. The difficulty in varying one dimensionless group to evaluate its effect upon $K_L a$, whilst keeping the remaining groups constant, represents one of the obstacles to a complete solution to this problem.

An alternative correlation for $K_L a$ has been proposed by Richards, J. W. (1961) by combining the equations of Rushton, J. H. (1951) for the mass transfer coefficient K_L and Calderbank, P. H. (1958) for the gas-liquid interfacial area a (equations (4.13) and (4.14) respectively):

$$\left(\frac{K_L D}{D_L}\right)\left(\frac{\mu}{\rho D_L}\right)^{a} = k'\left(\frac{D^2 N \rho}{\mu}\right)^{b} \tag{4.13}$$

$$a = c\left[\frac{(P/V)^{0\cdot4}\,\rho^{0\cdot2}}{\sigma^{0\cdot6}}\right]\left[\frac{V_s}{V_t}\right]^{0\cdot5} \tag{4.14}$$

where c = Constant
V_t = Terminal rising velocity of bubbles

By assuming constant liquid properties and a constant value of V_t and by assigning a value of 0·5 to the exponent b in equation (4.13) (Johnstone, R. E. and Thring, M. W., 1957), the two equations were combined and reduced to the following form

$$K_L a = K\left(\frac{P_g}{V}\right)^{0\cdot 4} V_s^{0\cdot 5} N^{0\cdot 5} \tag{4.15}$$

The above correlation for $K_L a$ is one alternative to that of Cooper, C. M. *et al.* (1944) (equation (4.3)); in fact, it gives a reasonable fit to the data of Cooper *et al.* The expression, like that of Cooper *et al.*, is not complete in that it does not include liquid properties. Furthermore, the data used by Calderbank in arriving at equation (4.14) were obtained with Newtonian fluids of low viscosity and consequently, it is likely that Richards' correlation is likewise confined to such fluids. Other correlations for $K_L a$ have been proposed by different workers, e.g. Yoshida, F. and Miura, Y. (1963). These different correlations for $K_L a$ have been successfully applied in certain circumstances; none have been applicable under all circumstances.

This summary of the most significant relationship established for aeration—agitation of non-viscous Newtonian fluids is not intended to be fully comprehensive. The reader is referred to several excellent reviews for this purpose, e.g. Finn, R. K. (1954), Richards, J. W. (1961), Sideman, S. *et al.* 1966), Wang, D. and Humphrey, A. E. (1968).

4.3 RHEOLOGY OF MOULD AND STREPTOMYCETE CULTURE FLUIDS

4.3.1 General Rheological Properties of non-Newtonian Fluids

When a force is applied to a liquid, it will normally flow; the viscosity of the liquid may be looked upon as its inherent resistance to flow.

Newton's law of viscous flow states that the tangential force (F) between two parallel layers of liquid of area A, distance δx apart and moving with a relative velocity δv is given by the following equation, when flow is not turbulent:

$$F = \mu A \frac{\delta v}{\delta x} \tag{4.16}$$

This equation may be re-arranged as follows

$$\mu = \frac{F/A}{\delta v/\delta x} = \frac{\tau}{\gamma} \tag{4.17}$$

where τ = Shear stress = F/A
γ = Shear rate = $\delta v/\delta x$

The shear stress is the applied force per unit area; the normal (c.g.s.) units are dynes/cm^2. The shear rate is the velocity gradient, the normal units being reciprocal seconds (sec^{-1}). The shear stress/shear rate ratio is the viscosity; the corresponding c.g.s. unit for viscosity is the poise. Thus, if we consider a centimetre cube of liquid (see Figure 4.5), when a tangential force of 1 dyne is applied to the upper surface and this surface moves with a velocity of

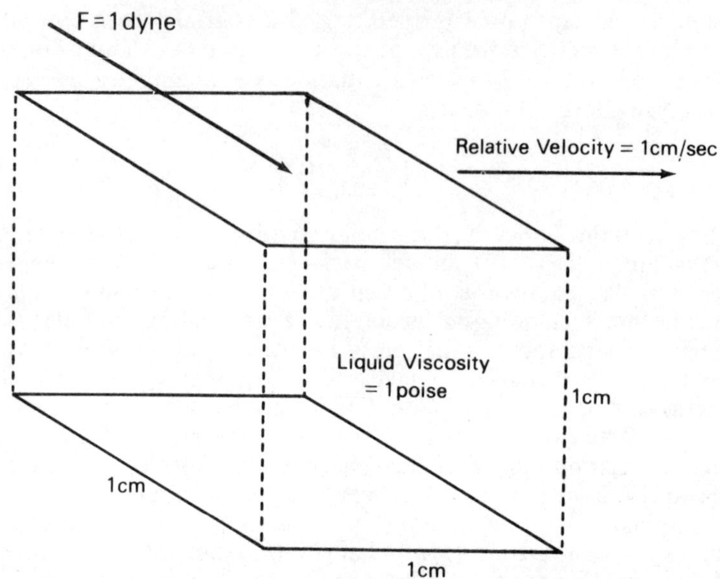

FIG. 4.5 Diagram illustrating the effect of the application of a tangential force of 1 dyne to a hypothetical cm cube of liquid, viscosity 1 poise.

1 cm/sec relative to the lower surface, the viscosity of the liquid will, by definition, be 1 poise. Simple solutions, e.g. salt and sugar solutions, obey Newton's law of viscous flow and are therefore said to be Newtonian in character. A plot of shear stress against shear rate on Cartesian co-ordinates is known as a rheogram, since it depicts the rheological properties of the fluid in question. A rheogram of a Newtonian fluid would be represented by a straight line, the slope of which would correspond to the viscosity of

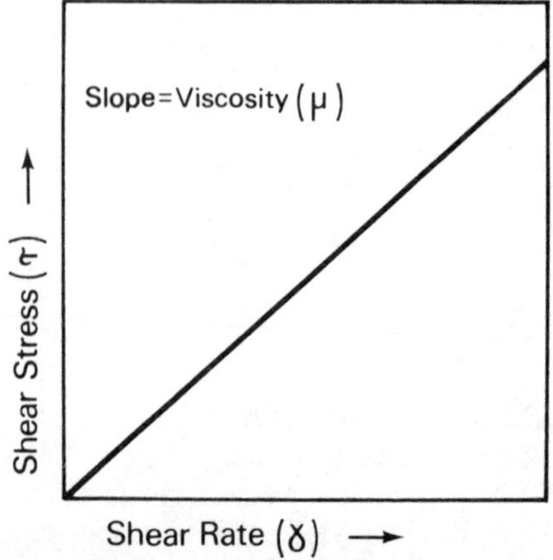

FIG. 4.6 Rheogram of a Newtonian liquid.

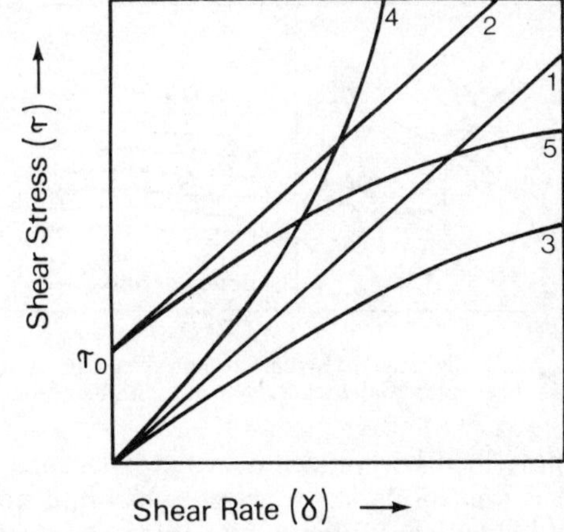

FIG. 4.7 Diagram illustrating rheological properties of different types of non-Newtonian liquids. 1, Newtonian; 2, Bingham plastic; 3, pseudoplastic; 4, dilatant; 5, Casson body.

the fluid (see Figure 4.6). The viscosity of a Newtonian fluid is thus a constant inherent property of that fluid. It will be obvious that, in order to determine the viscosity of a Newtonian fluid, it is necessary only to measure the shear stress at a single shear rate (or vice versa); a relatively simple viscometer (e.g. an Ostwald capillary viscometer) may be employed for this purpose.

Many fluids do not obey Newton's law of viscous flow and are therefore said to be non-Newtonian in character; such fluids do not possess a true viscosity but only an 'apparent viscosity' (μ_a), which is not a constant characteristic property of the fluid, being dependent upon the shear rate. It is not possible to comprehensively describe the rheological properties of all the different non-Newtonian fluids in an article of this type. However, some of the principal types of non-Newtonian fluids of interest to fermentation

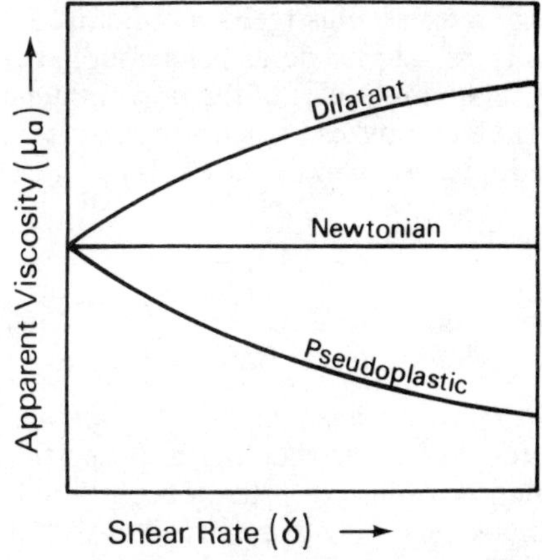

FIG. 4.8 Variation in apparent viscosity with shear rate.

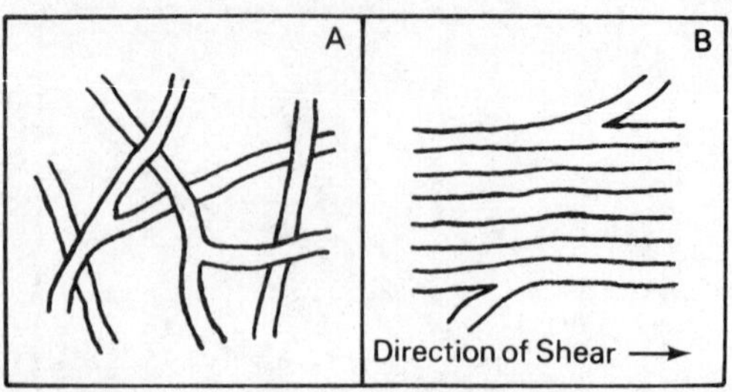

FIG. 4.9 Diagram illustrating alignment of hyphae at high shear rate. A, interwoven hyphae at low shear rate; B, aligned hyphae at high shear rate.

technologists are described here and illustrated by means of rheograms in Figure 4.7. Solutions and dilute suspensions, which are non-Newtonian in behaviour, may be divided conveniently into three main groups:

(i) *Bingham plastics.* These are similar to Newtonian fluids but differ from them in that there is a limiting shear stress (the yield stress or yield value) which must be exceeded if liquid flow is to occur. These liquids have a three-dimensional structure at rest of sufficient rigidity to withstand any stress less than the yield stress: once the yield stress is exceeded, the fluid then behaves in a Newtonian manner. The slope of the shear stress/shear rate plot is known as the 'coefficient of rigidity' or the 'plastic viscosity'. Thus both the terms yield stress and coefficient of rigidity are necessary to completely define the rheological properties of a Bingham plastic, which are represented by the following equation:

$$\tau = \tau_0 + \eta\gamma \tag{4.18}$$

where τ_0 = Yield stress
η = Coefficient of rigidity

(ii) *Pseudoplastics.* The characteristic property of a pseudoplastic is that the apparent viscosity (i.e. the slope of the shear stress/shear rate plot) decreases with increasing shear rate, as illustrated in Figures 4.7 and 4.8. Polymer solutions are generally pseudoplastic in behaviour; at progressively higher shear rates, the long-chain molecules of the polymer tend to align with each other and thus slip more readily, causing decreased viscosity (see Figure 4.9). The majority of pseudoplastics obey the power law, which is given in equation (4.9) below.

$$\tau = K(\gamma)^n \tag{4.19}$$

where K = Consistency coefficient
n = Flow behaviour index or power law index

The two terms, consistency coefficient and flow behaviour index are therefore necessary to completely define the rheological properties of a pseudoplastic fluid obeying the power law. The consistency coefficient has the same dimensions and units as viscosity; it may be regarded as the apparent viscosity, or shear stress, when the shear rate has a value of unity. The flow behaviour index is dimensionless and, for pseudoplastic fluids, its value is always less than unity. The smaller the value of the flow behaviour index, the greater the extent to which the rheological behaviour deviates from that of a Newtonian fluid.

(iii) *Dilatants.* The characteristic property of a dilatant fluid is that apparent viscosity increases with increasing shear rate (see Figure 4.8). The majority of dilatants also obey the power law, the value of the flow behaviour index always being greater than unity.

The following generalized equation may therefore be applied to the majority of fluids falling into the three categories described above:

$$\tau = \tau_0 + k'(\gamma)^n \tag{4.20}$$

where k' = constant (coefficient of rigidity for Bingham plastics, consistency coefficient for power law fluids).

The yield stress and flow behaviour index have the values indicated in Table 4.1 for each category.

Table 4.1 Characteristics of Different Types of Non-Newtonian Fluid

Category of fluid	τ_0	n
Newtonian	0	1
Bingham plastic	>0	1
Pseudoplastic	0	<1
Dilatant	0	>1

A fourth type of non-Newtonian fluid, known as a Casson body, was described by Casson in 1959; many concentrated suspensions behave as Casson bodies (Casson, N., 1959; Charm, S. E., 1963 and van Zanten, 1970). The characteristic features of a Casson body are that it possesses a well defined yield stress, whilst the apparent viscosity decreases with increasing shear rate (see Figure 4.7); it therefore combines the properties of a Bingham plastic and a pseudoplastic. However, the relationship between shear stress and shear rate is not governed by the generalized equation (4.20) previously given, but by the following relationship:

$$\sqrt{\tau} = \sqrt{\tau_0} + K_c\sqrt{\gamma} \tag{4.21}$$

where K_c = Casson viscosity.

In order to evaluate the properties of a given non-Newtonian fluid, it will be necessary to accumulate sufficient data to construct a rheogram. It will be evident that a relatively sophisticated viscometer, capable of meaning shear stress values over a wide range of shear rates, will be required for this purpose. Couette cup and bob, Brookfield open cup or cone and plate viscometers are the types frequently encountered in this application.

For Bingham plastics, the values of the yield stress and coefficient of rigidity can be simply determined directly from a rheogram. For pseudoplastics and dilatants obeying the power law, the values of the consistency

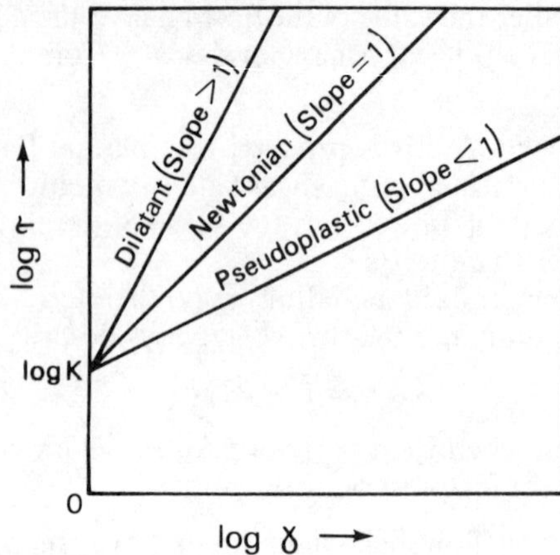

Fig. 4.10 Shear stress/shear rate plot on logarithmic co-ordinates for determination of flow behaviour index and consistency coefficient of power-law liquids.

coefficient and flow behaviour index may be calculated from viscometric data as follows. If equation (4.19) is converted to the logarithmic form, we obtain the following expression:

$$\log \tau = \log K + n \log \gamma \tag{4.22}$$

A plot of shear stress against shear rate on logarithmic co-ordinates will therefore yield a straight line, the slope of which is numerically equal to the value of the flow behaviour index (see Figure 4.10). Furthermore, the intercept on the shear stress axis (i.e. when shear rate has value of unity) is equal to the consistency coefficient. A log–log plot of this type will thus enable the values of both constants to be readily calculated. For a Casson body, a plot of $\sqrt{\tau}$ against $\sqrt{\gamma}$ on Cartesian co-ordinates (Figure 4.11) will obviously

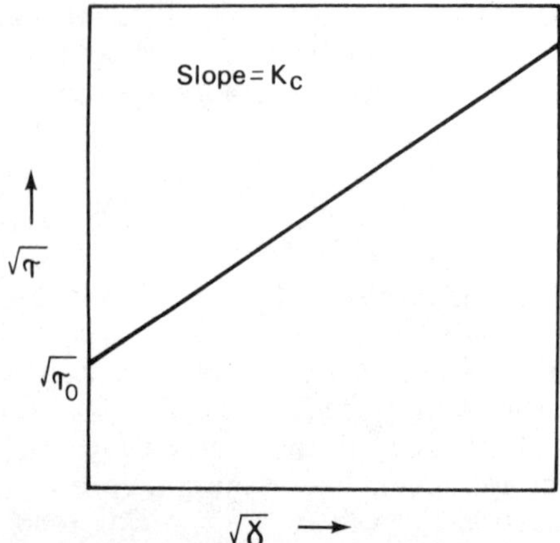

Fig. 4.11 Shear stress/shear rate plot for determination of Casson viscosity and yield stress.

yield a straight line whose slope is equal to the Casson viscosity, whilst the intercept gives the value of the yield stress.

4.3.2 Rheological Properties of Mycelial Suspensions

The presence of microbial cells suspended in a culture fluid will obviously considerably affect the rheological properties of that fluid.

Theoretically, one would expect Newtonian behaviour to be restricted to dilute suspensions of unicellular micro-organisms, i.e. to bacterial and yeast culture fluids. In such cases, the viscosity of the suspension will be dependent upon cell concentration and the size and shape of the individual cells. Einstein, A. (1906 and 1911) has shown that for dilute colloidal suspensions of spheres, the viscosity of the suspension is given by the following equation.

$$\mu_s = \mu_L (1 + 2{\cdot}5\phi) \tag{4.23}$$

where μ_s = Viscosity of suspension
μ_L = Viscosity of suspending liquid
ϕ = Fractional volume occupied by particles

Vand, V. (1948) has extended Einstein's equation to include particle interactions for more concentrated suspensions of spheres, as follows:

$$\mu_s = \mu_L(1 + 2{\cdot}5\phi + 7{\cdot}5\phi^2) \tag{4.24}$$

The latter equation is applicable to volume fractions as high as 0·25. Eirich, F. *et al.* (1936) conducted experiments to relate the concentration of suspensions of yeast cells and spores of the basidiomycete *Lycoperdon bovista* to the viscosity of the suspensions. Figure 4.12 illustrates how well their results correlate with the Einstein equation at low volume fractions and the Vand equation at high volume fractions. It seems likely that the equations of Einstein and Vand may well be applicable to a variety of culture fluids of unicellular micro-organisms.

Remarkably little has been published concerning the rheology of culture

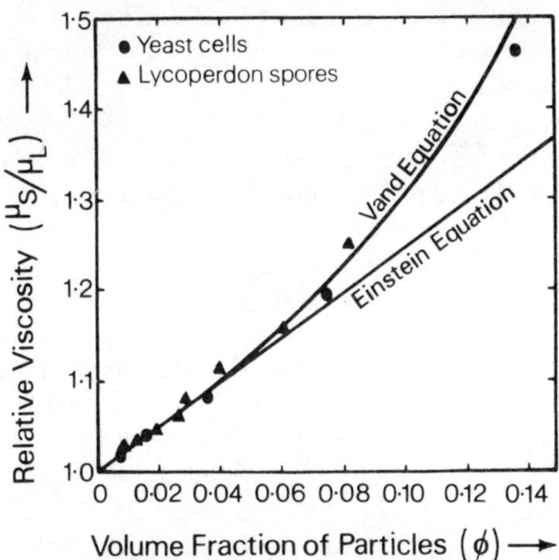

FIG. 4.12 Diagram illustrating effect of particle concentration upon suspension viscosity and applicability of Einstein and Vand equations. [*Eirich, F.* et al., *1936.*

fluids of filamentous micro-organisms such as moulds and streptomycetes, despite the obvious importance of this subject to fermentation technologists and biochemical engineers. Furthermore, much of the published information is incomplete and contradictory. The paucity of information on this subject is undoubtedly associated with the technical difficulties involved in obtaining reliable viscometric data on such heterogenous suspensions. Mycelial suspensions would be expected to exhibit non-Newtonian properties. The branched mycelial network forms a three-dimensional structure which imparts rigidity to the suspension; such suspensions might, therefore, be expected to exhibit a well defined yield stress. Also, the relatively long hyphae would be expected to align with each other at high shear rates, in much the same way as linear polymer molecules, with the result that the apparent viscosity would decrease with increasing shear rate. It might be anticipated, therefore, that mould and streptomycete culture fluids would behave as either Bingham plastics, pseudoplastics or Casson bodies. This has, indeed, proved to be the case: most of the published work indicates that mould or streptomycete suspensions behave rheologically as one of these three types of non-Newtonian fluid. Unfortunately no attempt has been made to relate rheological behaviour to the morphology of the organisms in submerged culture.

The mycelium of streptomycetes in submerged culture forms a dense mass of branched interwoven hyphae: the average diameter of the hyphae lies in the range 0·5–1·0 microns. Streptomycete culture fluids have been previously described as Newtonian, Bingham plastic and pseudoplastic in behaviour. Karow, E. O. *et al.* (1953) have reported that a culture of a streptomycin-producing strain of *Streptomyces griseus* behaved as a Bingham plastic. Deindoerfer, F. H. and West, J. M. (1960) reported that cultures of a nystatin-producing strain of *Streptomyces noursei*, taken at intervals throughout a fermentation, behaved, somewhat surprisingly, as a Newtonian fluid. Cultures of a streptomycin-producing strain of *S. griseus*, on the other hand, behaved either as a Newtonian fluid or as a Bingham plastic, depending on the age of the fermentation; Bingham plastic behaviour was evident approximately mid-way through the fermentation, whilst Newtonian characteristics were displayed during the initial and final stages of the fermentation. Newtonian behaviour in the terminal stage was attributed to marked fragmentation and lysis of the mycelium. Steel, R. and Maxon, W. D. (1962) reported that cultures of a novobiocin-producing strain of *Streptomyces niveus* behaved as a Bingham plastic, although they produced little evidence to support this statement. The culture fluids of two unidentified streptomycetes were examined rheologically by Tuffile, C. M. and Pinho, F. (1970); both behaved as Newtonian fluids during the initial stages of the fermentations and as pseudoplastics in the later stages.

The mycelium of moulds is similar to that of streptomycetes in submerged culture in that it forms a dense mass of branched inter-woven hyphae; the diameter of the hyphae is somewhat greater, however (2–5 microns). One of the earliest reports on the rheological properties of a mould culture fluid was that of Deindoerfer, F. H. and Gaden, E. L. (1955). They examined cultures of a penicillin-producing strain of *Penicillium chrysogenum* and found that they all behaved as Bingham plastics. However, Deindoerfer, F. H. and West, J. M. (1960) indicated that cultures of the same organism behaved as pseudoplastics; they explained this discrepancy by a failure to examine rheological behaviour at sufficiently low shear rates in the earlier work. The same workers found that cultures of the mould *Coniothyrium hellbori*, used for microbiological steroid hydroxylations, behaved as a Bingham plastic during the initial and final stages of the fermentation and as a pseudoplastic mid-way through the fermentation. Solomons, G. L. and Weston, E. O. (1961) have attributed Bingham plastic behaviour to cultures of an *Aspergillus sp.* whilst Taguchi, H. and Miyamoto, S. (1966) have ascribed pseudoplastic behaviour to a glucamylase-producing *Endomyces sp.* One of the most comprehensive investigations of the rheological behaviour of a fungal culture is that of Roels, J. A. *et al.* (1974). These workers employed cultures of *P. chrysogenum* for their investigations and found that the rheological characteristics of the cultures approximated more closely to those of a Casson body, rather than a pseudoplastic. They proposed a model for the rheological behaviour of filamentous suspensions, based upon an analogy with the theory of polymer physics (excluded volume concept), which included a so-called 'morphology factor'; this factor could be readily determined by experiment. They also produced evidence to support the view that their model might be applicable to other filamentous micro-organisms. An interesting feature of their work lies in the novel technique employed for evaluating the rheological properties of the culture fluids (Bongenaar, J. J. T. M. *et al.*, 1973) to which reference will be made later.

Thus, we see that accounts of the rheological properties of mould and streptomycete culture fluids are somewhat confused and apparently conflicting. Whilst different types of rheological behaviour undoubtedly exist in such cultures, it seems likely that much of the confusion in this field can be attributed to either or both of the following factors:

(i) Failure to correlate rheological behaviour with the morphology of the organisms in submerged culture and with the properties of the liquid (medium) in which the micro-organism is suspended.
(ii) The acknowledged technical difficulties in obtaining reliable viscometric data on heterogeneous mycelial suspensions.

With regard to the former factor, it will be obvious that, during the course of any fermentation, the cell density (mycelial weight) will increase and the morphological characteristics of the organism (degree of branching etc.) will alter; these changes will affect the rheological properties of the culture fluid. In the early stages of the fermentation, when little growth is evident, the organism will probably be distributed in the medium in the form of small discrete mycelial agglomerates; under such circumstances the culture fluid may well exhibit Newtonian characteristics (Tuffile, C. M. and Pinho, F., 1970). Later in the fermentation, when growth has been established and a three-dimensional mycelial network formed, the culture fluid will exhibit pronounced non-Newtonian behaviour. In the terminal stages of the fermentation, partial or complete lysis may occur; this will undoubtedly affect culture rheology. If partial lysis occurs, the effect will be unpredictable, at our present state of knowledge; if complete lysis occurs, the culture might be expected to revert to Newtonian behaviour (Deindoerfer, F. H. and West, J. M., 1960). The properties of the liquid medium in which the cells are suspended may be important if it contains polymers in solution which will contribute to the rheological behaviour of the mycelial suspension. Many media contain, for example, starch and unidentified polysaccharides extracted from materials such as soya bean meal or groundnut meal during medium sterilization; as the polysaccharide is degraded by the metabolic activities of the micro-organisms, its contribution to the rheological behaviour of the suspension will alter (Tuffile, C. M. and Pinho, F., 1970). Often polysaccharides are produced by the micro-organism itself: again, the presence of such polysaccharides will profoundly affect culture rheology (Banks, G. T. *et al.*, 1973; Leduy, A. *et al.*, 1974).

The technical difficulties in obtaining meaningful viscometric data relating to mycelial suspensions are considerable. With few exceptions, Couette cup and bob or Brookfield open cup viscometers have been employed for all the rheological studies previously mentioned. Such viscometers will give accurate or reproducible results with homogeneous polymer solutions; with heterogeneous mycelial suspensions, however, numerous problems are encountered. Some of the problems and errors associated with viscometric measurements in the presence of such suspensions are summarized below:

(*a*) Failure to obtain data over a sufficiently wide range of shear rates.
(*b*) Phase separation (i.e. of suspended solid from liquid) at the surface of the cylindrical measuring bob.
(*c*) Settling of the mycelium under the influence of gravity.
(*d*) The presence of dense mycelial aggregates (pellets) which impair measurement with narrow gap viscometers.
(*e*) The presence of trapped air bubbles in the suspension (easily overcome by evacuation).

The failure to obtain data over a sufficiently wide range of shear rates is worthy of further comment, since many of the less expensive viscometers will only permit measurements over a restricted range of shear rates. If, for

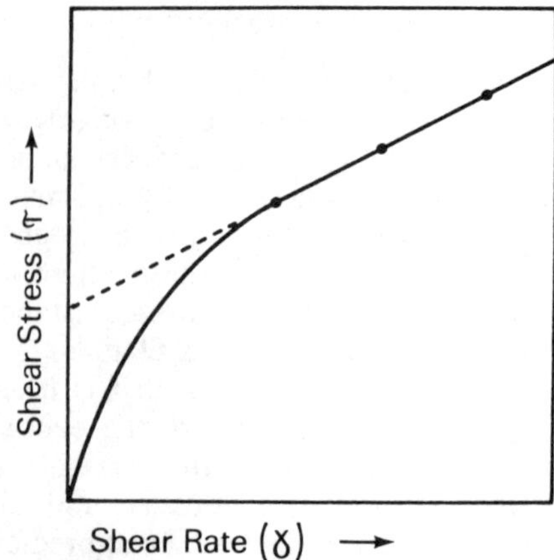

FIG. 4.13 Diagram illustrating mistaken characterization of a non-Newtonian fluid due to extrapolation of an inadequate number of measurements (see text).

example, measurements are omitted at low shear rates, a pseudoplastic fluid may often be mistaken for a Bingham plastic. This is exemplified in Figure 4.13; if the three data points are joined together by a straight line which is extrapolated to the shear stress axis (dotted line), plastic behaviour is suggested whereas measurements at lower shear rates would reveal pseudoplastic behaviour. An error of this type has been acknowledged by Deindoerfer, F. H. and West, J. M. (1960). It would seem likely that some of the earlier reports of plastic behaviour should be treated with reservation in view of this source of error. Some of the other technical problems associated with the measurement of viscosity of fermentation broths of filamentous micro-organisms have been discussed by Cheng, D. C. H. and Le Grys, E. A. (1975).

A very interesting approach to the problem of characterizing the rheological properties of cultures of filamentous micro-organisms lies in the technique of Bongenaar, J. J. T. M. *et al.* (1973) and used by Roels, J. A. *et al.* (1974) in their study of the rheological properties of cultures of *Penicillium chrysogenum*. Basically, the technique involves measuring the torque on a turbine impeller rotating in the culture fluid in an unbaffled cylindrical vessel of approximately three litres capacity under conditions of viscous or laminar flow, at different shaft speeds. The average shear stresses and shear rates in the system were calculated as shown below, thus enabling a rheogram of the culture fluid to be constructed. The average shear rate has been shown (Metzner, A. B. and Otto, R. E., 1957; Calderbank, P. H. and Moo-Young, M. B., 1959) to be directly related to the impeller rotational speed by the following equation:

$$\bar{\gamma} = kN \tag{4.25}$$

where $\bar{\gamma}$ = Average shear rate
k = Proportionality factor or constant

The average shear stress in the system was derived as follows. Each of the two well known and established expressions shown below may be re-arranged in the manner indicated

(a) Normal viscosity equation

$$\bar{\mu}_a = \frac{\bar{\tau}}{\bar{\gamma}} \rightarrow \bar{\tau} = \bar{\mu}_a \bar{\gamma} \tag{4.26}$$

where $\bar{\mu}_a$ = Average apparent viscosity
$\bar{\tau}$ = Average shear stress

(b) Rushton's equation for power consumption in the region of viscous flow

$$P = c\bar{\mu}_a N^2 D^3 \rightarrow \bar{\mu}_a = \frac{P}{cN^2D^3} \tag{4.27}$$

From equations (4.25–4.27) it follows that

$$\bar{\tau} = \frac{Pk}{cND^3} \tag{4.28}$$

However,

$$P = 2\pi NT \tag{4.29}$$

where T = Torque

Combination of equations (4.28 and 4.29) yields the following expression

$$\bar{\tau} = \frac{2\pi Tk}{cD^3} \tag{4.30}$$

The work of Metzner, A. B. and Otto, R. E. (1957) and Calderbank, P. H. and Moo-Young, M. B. (1959) has indicated that the proportionality factor k has a value of approximately 11·5 over a wide range of different conditions; furthermore, the geometry dependent constant c in Rushton's equation for power consumption has been shown by numerous workers, including Rushton himself, to have a value of approximately 64. Equation (4.30) thus reduces to:

$$\bar{\tau} = \frac{1{\cdot}13T}{D^3} \tag{4.31}$$

The average shear stress is thus easily calculated from the torque exerted on the agitator shaft. This technique for evaluating the rheological properties of culture fluids of filamentous organisms is less fundamental than conventional viscometric techniques, since shear stress and shear rate are not measured directly under rigorously defined conditions. The theoretical basis of the method rests upon the concept of average shear rate and average apparent viscosity in stirred vessels introduced by Metzner and Otto, to which more detailed reference will be made later, and the power consumption relationship established by Rushton. However, the technique has the considerable advantage that it avoids many of the technical difficulties mentioned earlier associated with conventional viscometric measurements on such fluids and therefore gives very reproducible results. The only limitation of the technique is that, since all measurements are conducted under laminar flow conditions, the range of shear rates involved is therefore restricted and may not be relevant to those encountered under actual fermentation conditions.

It is obvious that one major gap in our knowledge and understanding of aeration-agitation of mould and streptomycete cultures concerns the lack of information available on the rheological behaviour of such cultures. Further study in this field would be rewarding. In particular, the technical difficulties involved in performing viscometric measurements must be overcome or avoided whilst attempts should be made to relate rheological behaviour to the morphological characteristics of the micro-organism.

4.4 AERATION-AGITATION OF VISCOUS NON-NEWTONIAN FLUIDS

4.4.1 General Problems

In attempting to reach an understanding of aeration-agitation phenomena in the presence of viscous non-Newtonian fluids such as mould and streptomycete cultures, it is desirable to begin by clearly identifying the nature of the difficulties involved. These are briefly described below:

(i) It is normal practice, in the field of fermentation technology, to endeavour to work in the region of turbulent flow, where the relationships between operating variables (e.g. impeller shaft speed, impeller diameter, air flow rate, liquid density and viscosity), power consumption and aeration efficiency are well understood. Unfortunately, the high viscosity of the majority of mould and streptomycete cultures prevents conditions of fully developed turbulent flow from being established; with such cultures, the flow conditions within the fermenter usually lie in the extensive transition region between the regions of viscous and turbulent flow. In the transition region, the relationships between operating variables and power consumption are complex, correlation necessitating the construction of a complete power curve for the fermenter in question, whilst little is known concerning the relationship between power consumption and aeration efficiency.

(ii) Non-Newtonian fluids, such as mould and streptomycete culture fluids, do not possess a true viscosity but only an apparent viscosity which is not a constant, being dependent upon the shear rate. It will be obvious that different shear rates exist in different parts of a fermenter, the maximum shear rate being in the immediate vicinity of the impeller. It has, in fact, been shown that shear rate decreases exponentially with distance from the impeller tip (Otto, R. E., 1957; Taylor, J. S., 1955). Since the shear rate varies within the fermenter, the apparent viscosity, being dependent upon it, will likewise vary. Thus a complex system, with regard to liquid flow and properties, exists within the fermenter which has, in the past, defied theoretical analysis. The inability to ascribe a value to the viscosity of the fermenter contents has hindered progress in this field, bearing in mind that, without such an ascribed value, it is impossible to calculate Reynold's Numbers and hence construct power curves.

(iii) The third major difficulty is the one to which reference has already been made in the preceding section, namely the lack of information available concerning the rheological behaviour of culture fluids of filamentous micro-organisms.

4.4.2 Power Consumption

Un-gassed fluids. A way out of the second of the major difficulties mentioned above, namely the complex flow behaviour and liquid properties within the fermenter or reactor, was suggested in the pioneering work of Metzner, A. B. and Otto, R. E. in 1957. They introduced the concept of the *average* shear rate in an agitated vessel and proposed that it was related to the rotational speed of the impeller by a simple proportionality equation:

$$\bar{\gamma} = kN \qquad (4.25)$$

This is an extremely useful equation for design purposes. Its significance lies in the fact that it can be used to calculate the *average* apparent viscosity of the vessel contents if the value of the proportionality factor is known together with the relationship between shear rate and apparent viscosity for the fluid concerned. The relationship between shear rate and apparent viscosity can be ascertained from normal viscometric measurements; for a pseudoplastic liquid obeying the power law, as used by Metzner and Otto, the following expression for average apparent viscosity can be obtained from equations (4.17) and (4.19)

$$\bar{\mu}_a = K(\bar{\gamma})^{n-1} \tag{4.32}$$

or, substituting for $\bar{\gamma}$

$$\bar{\mu}_a = K(kN)^{n-1} \tag{4.33}$$

Metzner and Otto employed a method, previously published by Magnusson, K. (1952), for calculating the average apparent viscosities of non-Newtonian fluids in agitated vessels from the power curve for Newtonian liquids. From the results so obtained, they were able to determine values for the proportionality factor k in equation (4.25). The method used by Metzner and Otto to determine the value of the proportionality factor is briefly described below:

(a) A power curve was constructed for a Newtonian fluid in the selected vessel, in the normal manner.
(b) Power consumption was measured over a range of different agitator shaft speeds, using a non-Newtonian fluid in the same vessel.
(c) Using the data obtained in (b), the value of the Power Number was calculated at each agitator speed for the non-Newtonian fluid ($N_p = P/\rho N^3 D^5$).
(d) The Reynolds Numbers corresponding to the Power Numbers calculated in (c) for the non-Newtonian fluid, were obtained by interpolation of the power curve for the Newtonian fluid, *in the region of viscous flow only*. It was assumed that the power curves for the two fluids must be identical in this region, since viscous forces alone determine power consumption.
(e) The average apparent viscosity of the non-Newtonian fluid at each agitator speed was calculated from the normal Reynolds Number equation ($N_{Re} = \rho N D^2/\bar{\mu}_a$ or $\bar{\mu}_a = \rho N D^2/N_{Re}$.
(f) From the data obtained (e), a log–log plot of average apparent viscosity against agitator shaft speed was constructed.
(g) A similar log–log plot of apparent viscosity against shear rate was constructed from normal viscometric measurements on the non-Newtonian fluid. (It will be seen that if equation (4.32) is converted to the logarithmic form, a log–log plot of μ_a against γ will yield a straight line of slope $(n-1)$ (see Figure 4.14).
(h) From the log–log plots constructed in (f) and (g), values of the shear rate and agitator speed were selected at identical viscosities and were subsequently plotted against each other on Cartesian co-ordinates. The slope of this plot, which proved to be linear, is equal to the value of the proportionality factor (from equation

In their studies, Metzner and Otto used pseudoplastic polymer solutions in conventional baffled cylindrical vessels, each with a single flat bladed turbine impeller. They found, using the method outlined above, that the proportionality factor, relating average shear rate and impeller shaft speed, had a value of approximately 13 in their experiments.

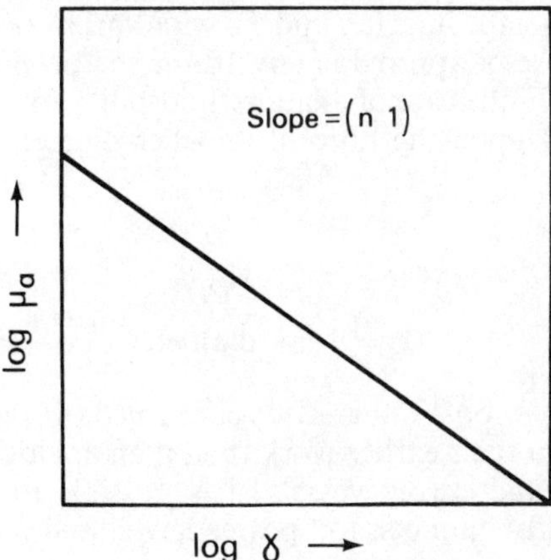

FIG. 4.14 Apparent viscosity/shear rate plot on logarithmic co-ordinates for a pseudoplastic, power-law fluid.

Using the value of the proportionality factor so obtained, Metzner and Otto were able to calculate Reynolds Numbers outside the region of viscous flow for the non-Newtonian fluids and thus extend the power curve for these fluids. They showed that the region of viscous flow for a non-Newtonian fluid extended beyond that for a Newtonian fluid (see Figure 4.15).

Confirmation of Metzner and Otto's work was quick to follow. In 1959, Calderbank, P. H. and Moo-Young, M. B. reported data for Bingham plastic, pseudoplastic and dilatant fluids in a vessel equipped with a single impeller of variable design and size. For Bingham plastics and pseudoplastics, they clearly demonstrated a linear relationship between average shear rate and

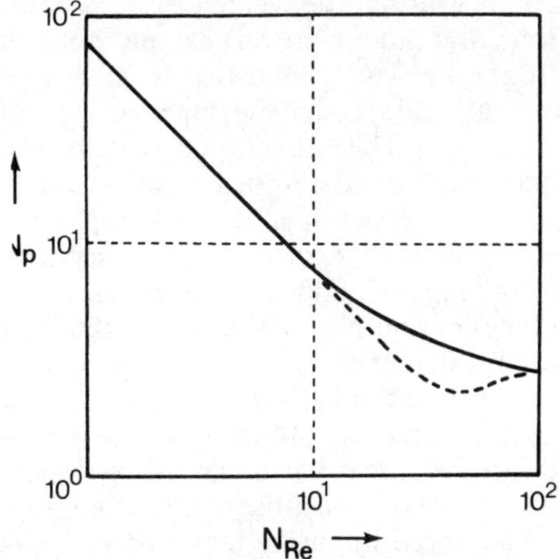

FIG. 4.15 Pseudoplastic deviation from Newtonian power curve. Discontinuous line refers to pseudoplastic liquid. [*Metzner, A. B. and Otto, R. E., 1957.*

the rotational speed of the impeller under a wide variety of different conditions and obtained a value of approximately 10 for the proportionality factor of Metzner and Otto. In the case of dilatant fluids, they found that the relationship was dependent upon the ratio of impeller diameter to tank diameter, thus:

$$\bar{\gamma} = 12{\cdot}8\, N\left(\frac{D}{D_t}\right)^{0\cdot 5} \tag{4.34}$$

where D_t = Tank diameter

Calderbank and Moo-Young also confirmed a suggestion made by Metzner and Otto in their earlier work that, if an analogy is drawn between fluid flow in pipes and stirred vessels, it is possible to obtain a modified, generalized Reynolds Number for power-law liquids, represented by the following equation:

$$N_{Re} = \frac{D^2 N\rho(8N)^{1-n}}{K}\left(\frac{4n}{3n+1}\right)^n \tag{4.35}$$

The corresponding equation for average apparent viscosity is:

$$\bar{\mu}_a = \frac{K}{(8N)^{1-n}}\left(\frac{3n+1}{4n}\right)^n \tag{4.36}$$

They demonstrated a good correlation between the average apparent viscosity calculated from the above equation and that obtained by the procedure of Metzner and Otto.

Further confirmation of the linear relationship between average shear rate and impeller rotational speed was provided in 1960 by Metzner, A. B. and Taylor, J. S. These workers measured local shear rates directly in a vessel in the presence of a viscous Newtonian and pseudoplastic liquid. Their technique involved measuring the velocities of small Plexiglas spheres suspended in the fluid and moving in vertical and horizontal planes of light by time-lapse photography. They found that local shear rates were directly proportional to the rotational speed of the impeller, as predicted by the earlier work of Metzner and Otto. They also found that the shear rate decreased more rapidly with distance from the impeller in the case of the pseudoplastic fluid than in the case of the Newtonian fluid, as might be expected.

In 1961, Metzner, A. B. *et al.* extended the earlier work of Metzner and Otto to cover a wider range of conditions. Experimental variables included the use of different types of non-Newtonian fluids (Bingham plastics, pseudoplastics and dilatants), different designs and sizes of impeller, different vessel dimensions, use of baffled and unbaffled vessels and, in particular, the use of multiple impeller systems and high impeller diameter/tank diameter ratios; the two last-named variables are of particular interest since multiple impellers are frequently used in industry, whilst the relationship between average shear rate and impeller shaft speed might be expected to break down as the tip of the impeller approaches the wall of the vessel. The relationship was found to apply under all conditions tested, except in the case of dilatant fluids where the flow behaviour index had a high value or the impeller diameter/tank diameter ratio approached unity. The values of the proportionality factor determined by them lay in the range 10–13; they suggested that a general value of 11 could be used for calculation purposes with very little error. Also, in 1961, Calderbank, P. H. and Moo-Young, M. B. produced a general correlation for the power required to agitate both Newtonian and non-Newtonian fluids with agitators of various designs. The correlation made use of the generalized Reynolds Number mentioned earlier (equation (4.35)). and allowed power consumption to be predicted in the regions of viscous and turbulent flow, without recourse to previously obtained experimental data, from specified impeller/vessel dimensions and fluid properties. Values in the range 10·0–11·6 were found for the proportionality factor.

In the studies described above, the pseudoplastic (and dilatant) liquids employed all obeyed the power law relating shear stress to shear rate. Godleski, E. S. and Smith, J. C. (1962) demonstrated that the relationship between average shear rate and impeller rotational speed also held for pseudoplastic liquids which did not obey the power law, i.e. where the value of the flow behaviour index was not constant but dependent on the shear rate; in these studies the proportionality factor was found to have a value of approximately 11.

It will thus be seen that the original proposal of Metzner and Otto that the average shear rate in a stirred vessel is directly proportional to the rotational speed of the impeller has been amply confirmed under a wide variety of different conditions. In fact, this simple relationship has been shown to be substantially independent of impeller/vessel geometry and the rheological properties of the fluid. The importance of Metzner and Otto's relationship cannot be over-emphasized; by enabling the average shear rate in the vessel to be simply calculated, and hence the average apparent viscosity (provided the necessary viscometric data are available), the power required for agitation can be determined from a Reynolds Number–Power Number plot (power curve) in the same manner employed for Newtonian fluids.

An interesting fact to emerge with clarity from some of the studies (Metzner, A. B. and Otto, R. E. 1957; Metzner, A. B. *et al.*, 1961; Godleski, E. S. and Smith, J. C., 1962) was the greater difficulty experienced in achieving adequate bulk mixing with pseudoplastic rather than Newtonian fluids; considerably greater power consumption was required to achieve the same degree of bulk mixing in the case of the non-Newtonian fluids. This phenomenon is to be expected, however, in view of the fact that shear rate decreases more rapidly with distance from the impeller in the case of pseudoplastic fluids. Another, less obvious, observation was made by Godleski and Smith; they found that bulk mixing of pseudoplastic fluids was most rapid in un-baffled vessels at agitator speeds which caused a vortex to form. The mixing time, or blend time, was inversely proportional to the square root of the vortex depth:

$$t_m = k' \, r \, H_v^{-0{\cdot}5} \tag{4.37}$$

where r = Impeller radius
H_v = Depth of vortex

The non-Newtonian fluids employed in the foregoing studies were polymer solutions or mineral suspensions, i.e. non-biological systems. Very little information is available in the literature concerning the applicability of Metzner and Otto's relationship to the non-Newtonian culture fluids of filamentous micro-organisms. Although there is no apparent reason to believe that the relationship would not apply to such fermentation broths, this lack of information represents a gap in our knowledge which needs to be remedied. Taguchi, H. and Miyamoto, S. (1966) have, however, produced power consumption data for an ungassed pseudoplastic Endomyces culture, used for glucamylase production, which imply the validity of Metzner and Otto's relationship under such conditions.

Gassed fluids. Little information is available concerning the effect of gassing upon power consumption in the presence of viscous non-Newtonian fluids. Obviously, gassing will decrease power consumption, as in the case of Newtonian fluids, but comprehensive correlations for gassed power consumption with the former type of fluid have not been established. The most widely used correlation for Newtonian fluids, as mentioned previously, is that of Michel, P. J. and Miller, S. A. (1962) (equation (4.10)).

One of the few investigations in this area, involving non-Newtonian fluids, is that of Taguchi, H. and Miyamoto, S. (1966). These workers studied gassed power consumption using various pseudoplastic glucamylase producing *Endomyces* cultures. They plotted their data as the ratio of gassed power to ungassed power (Pg/Po) against the Aeration Number in the manner of Oyama, Y. and Endoh, K. (1955). However, unlike Oyama and Endoh, they found that their results, presented in this form, were dependent on both the Reynolds Number and the rheological properties of the fluid (Figure 4.16). In this respect, it is interesting to note that in an earlier work Nishikawa, F. (1965), using pseudoplastic cartoxymethylcellulose solutions, showed that both bubble velocity and gas hold-up were related to the rheological properties of the fluids and, in particular, to the flow behaviour index. Taguchi and Miyamoto also found that the correlation of Michel and Miller applied to their non-Newtonian *Endomyces* cultures, but only in the

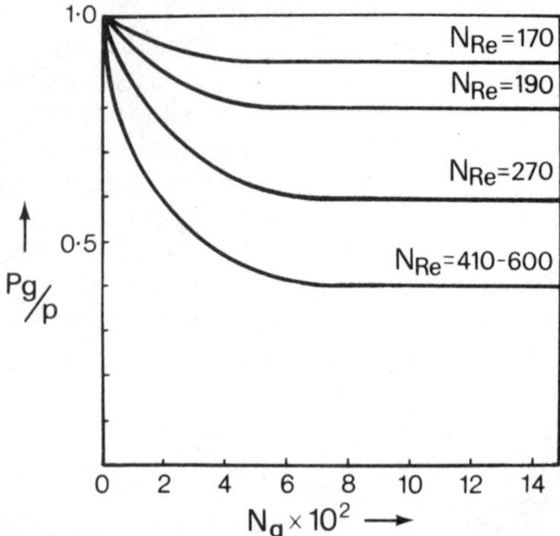

FIG. 4.16 Relationship between gassed power consumption and aeration number for pesudoplastic endomyces culture. [*Taguchi, H. and Miyamoto, S., 1966.*

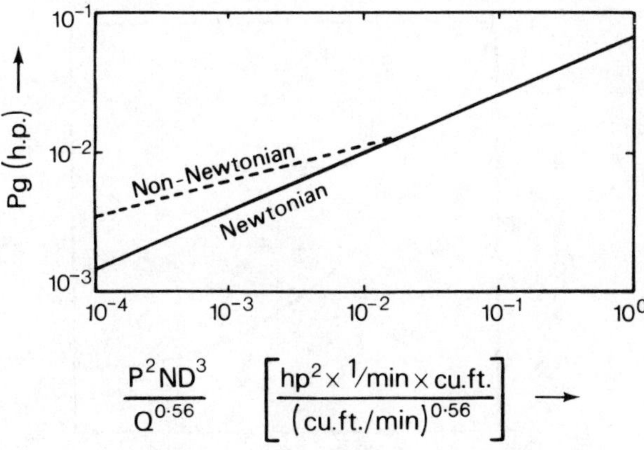

FIG. 4.17 Applicability of Michel and Miller correlation for gassed power consumption to pseudoplastic endomyces culture. [*Taguchi, H. and Miyamoto, S., 1966.*

region of turbulent flow (Figure 4.17). Since conditions of fully developed turbulent flow seldom exist in viscous mould and streptomycete culture fluids, this discovery would appear to be of limited practical value. Once again, there is ample scope for investigation in this area.

A fact, which is not generally appreciated, is that gassed power consumption actually **decreases** during the course of a typical mould or streptomycete fermentation. It is wrongly assumed by many scientists and technologists that the increasing viscosity of the culture, due to growth of the micro-organism, causes power consumption to increase. The explanation of this phenomenon is quite simple. In the early stages of the fermentation, when culture viscosity is low, the air is well distributed throughout the culture fluid in the form of small bubbles by the agitation system; power consumption is therefore high. As the culture viscosity increases, the air becomes less well distributed and, in fact, tends to rise up the centre of the fermenter in the form of large bubbles; the presence of these large bubbles in the immediate vicinity of the impeller, where most of the power is absorbed, lowers the average density of the gas/liquid mixture and hence the power consumption. The decrease in gassed power consumption experienced during such fermentations, due to increased culture viscosity, can be considerable; for example, power consumption can be reduced to as little as 30% of its initial value. This adverse effect of viscosity upon air distribution is undoubtedly one of the major factors responsible for the decrease in aeration efficiency which also accompanies growth of the micro-organism.

4.4.3 Aeration Efficiency

General considerations. The presence of the mycelium of filamentous fungi and streptomycetes in submerged culture has a profound effect upon oxygen transfer and aeration efficiency. This fact has been clearly demonstrated by a number of different investigators, e.g. Deindoerfer, F. H. and Gaden, E. L. (1955), Brierley, M. R. and Steel, R. (1958), Solomons, G. L. and Weston,

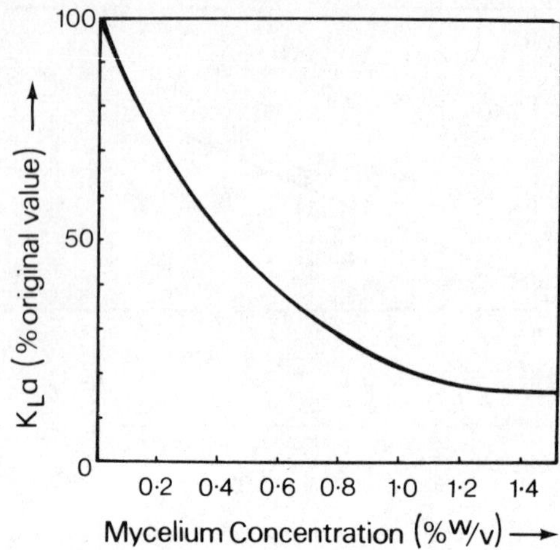

FIG. 4.18 Effect of mycelium of penicillium chrysogenum upon aeration efficiency. [*Deindoerfer, F. H. and Gaden, E. L., 1955.*

E. O. (1961), Steel, R. and Maxon, W. D. (1962) and others. A graph representing a typical relationship between aeration efficiency and mycelium concentration is shown in Figure 4.18; this graph is taken from the work of Deindoerfer, F. H. and Gaden, E. L. (1955) who used mycelium of the mould *Penicillium chrysogenum.*

As will be seen from Figure 4.18, the reduction in aeration efficiency, caused by the presence of mycelium, can be very appreciable. For example, Deindoerfer and Gaden found that the volumetric mass transfer coefficient (K_La)

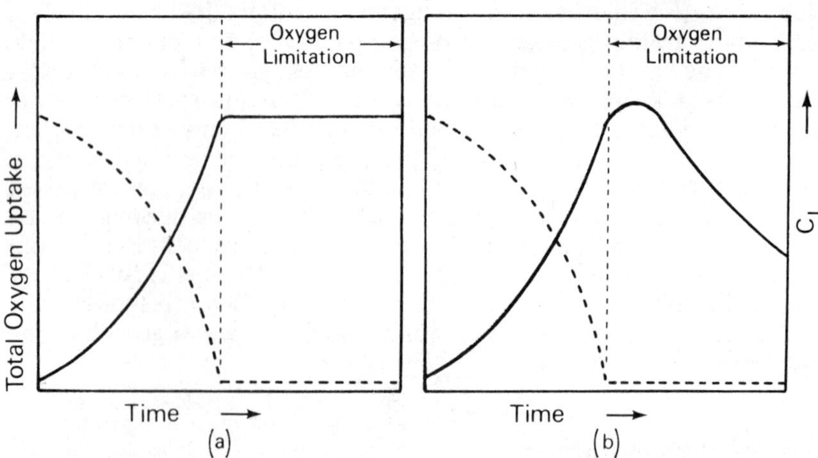

FIG. 4.19 Course of events during oxygen limited fermentations.

(a) Typical bacterial fermentation.
(b) Typical fungal fermentation.
__________ Total oxygen uptake rate of culture.
\- - - - - - - Dissolved oxygen concentration.

could be reduced by 85% at a mycelial concentration of 1·35% (dry weight) whilst Brierly, M. R. and Steel, R. (1958) demonstrated a 90% reduction in K_La due to the presence of *Aspergillus niger* mycelium at a concentration of 2% (dry weight).

A very common example of the effect of mycelium upon aeration efficiency, which will be familiar to all who work routinely with large-scale mould or streptomycete fermentations, lies in the characteristically different oxygen uptake patterns which result from such fermentations and those involving unicellular micro-organisms, such as bacteria and yeasts. This difference is exemplified in Figures 4.19a and 4.19b which represent respectively the course of events in a typical oxygen-limited bacterial fermentation and a similar mould fermentation.

It will be seen that in both cases, up to the time of oxygen limitation (indicated by the dotted lines), approximately exponential growth occurs; the dissolved oxygen concentration decreases until it reaches the critical level at the time of oxygen limitation. Thereafter the pattern of events is quite different in each of the two cases. In the case of the bacterial fermentation, the total oxygen uptake reaches a maximum value at the time of oxygen limitation and remains at this maximum value until some other nutrient becomes limiting. During this period of high constant oxygen uptake, the oxygen transfer rate is limited by the equipment employed and linear growth results. In the case of the mould fermentation, however, the total oxygen uptake rate **decreases** during the period of oxygen limitation; this decrease can only be attributed to the increasing viscosity of the culture, caused by the increasing mycelial concentration. These two quite different courses of events, characteristic of oxygen-limited fermentations of unicellular and multicellular filamentous micro-organisms, have been repeatedly observed during pilot-scale fermentations conducted at this establishment.

Although a certain amount of quantitative information is available in the scientific literature, concerning aeration efficiency in the presence of viscous non-Newtonian systems, few attempts have been made to correlate aeration efficiency (K_La) with operating variables under these conditions. It will be obvious that, if any correlation is to be complete, it must include the relevant physical properties of the fluid concerned such as density, viscosity, surface tension and the diffusivity of oxygen in the liquid. Viscosity and surface tension are probably the most important of the liquid properties mentioned, since they can vary considerably between different types of fermentation and during the course of a single fermentation. Density and oxygen diffusivity would not be expected to vary to anything like the same extent in such systems. Surface tension, which has not been discussed in any detail until now, will obviously influence aeration efficiency by affecting both the size of the gas bubbles formed and their rate of movement in the liquid (and hence gas hold-up). The motion of gas bubbles in non-Newtonian liquids has been reviewed by Astarita, E. and Apuzzo, E. (1965). Some of the more important studies and attempted correlations, concerning oxygen transfer in the presence of viscous non-Newtonian systems, are described in the next two sections.

Correlations between K_La *and power consumption.* As mentioned earlier in this article, the most well-known and widely accepted correlation for aeration efficiency, with non-viscous Newtonian systems, in the region of turbulent flow, is that with power consumption. Strict proportionality between the volumetric transfer coefficient and power consumption is often assumed in prediction of results and scaling-up procedures. It is not surprising therefore, that attempts have been made to establish similar correlations for viscous non-Newtonian fluids.

A very interesting series of investigations, which throw a certain amount of light on the subject, are those of Maxon, W. D. and Steel, R.; throughout their investigations they worked with cultures of a novobiocin-producing strain of *Streptomyces niveus* which, according to them, exhibited Bingham plastic behaviour. The starting-point in their studies was the observation by Maxon, W. D. (1959) that in 20 litre scale fermentations, greater power input was required to achieve a given yield of novobiocin with a large impeller (49% of tank diameter) than with a smaller impeller. This phenomenon was further explored by Steel, R. and Maxon, W. D. (1962) again using a 20 litre

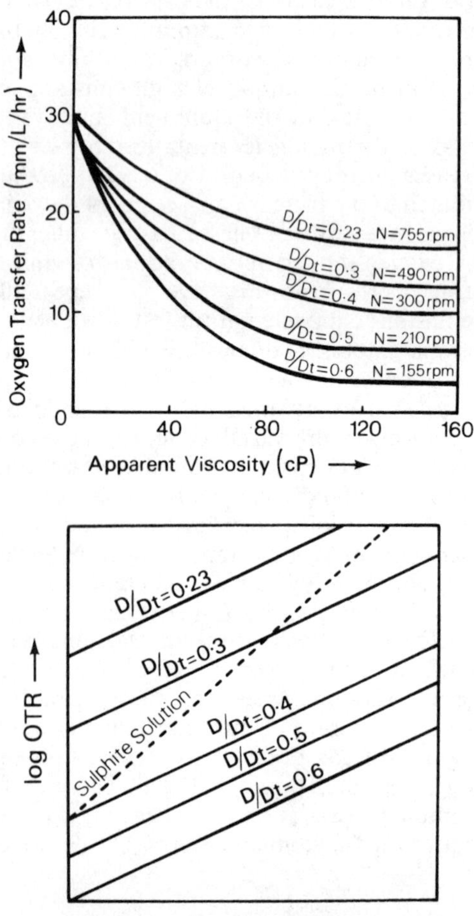

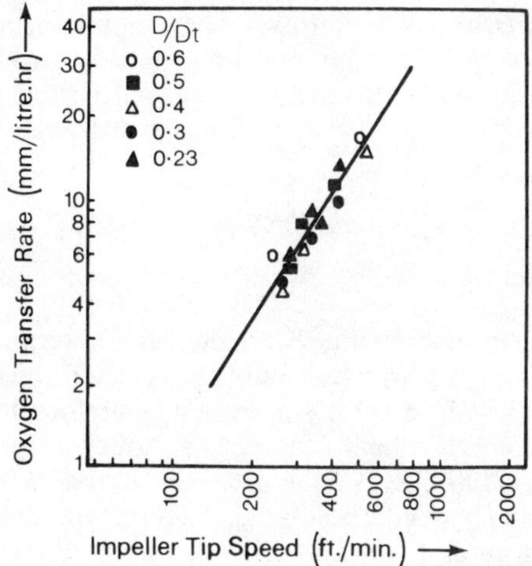

FIG. 4.20 Diagram illustrating experiments of Steel, R. and Maxon, W. D. (1962), using cultures of a novobiocin-producing strain of *Streptomyces niveus.*

(a) Effect of apparent viscosity on oxygen transfer rate at different impeller rotational speeds and impeller diameter/tank diameter ratios.
(b) Effect of power consumption on oxygen transfer rate at different impeller diameter/tank diameter ratios.
(c) Correlation between oxygen transfer rate and impeller tip speed.

fermenter with a range of geometrically similar turbine impellers of different diameters. Initially, they investigated the effect of fluid viscosity on oxygen transfer rate with the different impeller systems. (It should be noted that the fluid viscosity quoted by them was the apparent viscosity, measured at a pre-determined shear stress in a Stormer viscometer.) The effect of increasing fluid viscosity on oxygen transfer rate at constant rotational impeller speed was examined for each impeller diameter; the agitator speeds were adjusted in each case to give the same power consumption in water. They found, as expected, that the oxygen transfer rate decreased with increasing viscosity; however, the decrease was proportionally greater for larger diameter impellers (see Figure 4.20a). Power consumption as a function impeller rotational speed was also measured for each size of impeller. A plot of oxygen transfer rate against power consumption on logarithmic co-ordinates for each impeller system, using a high viscosity culture fluid, is reproduced in Figure 4.20b. It will be seen that, for each impeller diameter, there was a clear relationship between oxygen transfer rate and power consumption ($K_La = k\,Pg^{0\cdot46}$). However, the oxygen transfer rate achieved at a given power input was markedly dependent on the impeller diameter. Smaller impellers gave better results for a given power input; for example, it was found that the smallest impeller ($D/Dt = 0{\cdot}23$) gave an oxygen transfer rate eight times higher than the largest impeller ($D/Dt = 0{\cdot}60$) at equal power input. A point of interest made by Steel and Maxon was that the small impellers were capable of transferring oxygen to this viscous non-Newtonian culture at a rate higher than to aqueous sodium sulphite solution (a non-viscous Newtonian system). Since they had failed to correlate oxygen transfer rate with power consumption independently of impeller diameter, Steel and Maxon examined their results for other possible correlations. They found, in fact, that they were able to correlate oxygen transfer rate quite satisfactorily with impeller tip speed. Their relationship between oxygen transfer rate and impeller tip speed is illustrated in Figure 4.20c and is consistent with the following equation:

$$OSR = k'(S)^{1 \cdot 6} = k'(\pi ND)^{1 \cdot 6} \tag{4.38}$$

where S = Impeller tip speed

In order to explain their results, Steel and Maxon proposed that in viscous non-Newtonian systems the gas bubbles, once formed, do not readily coalesce. In such systems it is only necessary, therefore, to provide a zone of high shear (the maximum shear rate is at the impeller tip) to effect the initial formation of small bubbles. In these circumstances, it might be expected that aeration efficiency would be a function of impeller tip speed, rather than power consumption, as Steel and Maxon observed. By contrast, in non-viscous Newtonian systems, there is a well-known tendency for gas bubbles to coalesce. In this case, it would be necessary to provide sufficient power not only to disperse the air in the form of small bubbles at the impeller tip but also to maintain small bubbles within the system, i.e. to prevent or minimize bubble coalescence; greater power input would be required for this purpose and aeration efficiency might be expected to correlate with power consumption rather than impeller tip speed under these circumstances. As evidence to support their theory, Steel and Maxon drew analogies with other systems. For example, Fondy, P. L. and Bates, R. L. (1961) found that the particle size of a sodium-potassium alloy sheared in mineral oil correlated with impeller tip speed; particle size varied with impeller tip speed raised to the power $-1{\cdot}8$ (c.f. equation (4.38)). Also, Finn, R. K. (1960) noted a similar correlation between oil droplet size in sheared suspensions and impeller tip speed, provided that the suspensions were stabilized with a surface active agent. In both of the quoted analogous cases, it is important that recoalescence is prevented; in the latter case, if the surface active agent was omitted, oil droplet coalescence occurred and droplet size no longer correlated with impeller tip speed but with power consumption instead.

Steel and Maxon continued their studies with *S. niveus* cultures (Steel, R. and Maxon, W. D., 1966a) by examining the effects of impeller rotational speed and impeller diameter upon dissolved oxygen concentrations in different parts of the fermenter and upon total oxygen uptake on the 20, 250 and 15,000 litre scales of operation. They demonstrated that, under certain conditions (notably small impeller diameter/tank diameter ratios), different dissolved oxygen concentrations occurred in different parts of the fermenter, a clear indication of inadequate bulk mixing. The effects of this poor bulk mixing were difficult to interpret, although Steel and Maxon assumed that poor mixing contributed to the maintenance of oxygen-limiting conditions. An even more interesting observation was that, even with good bulk mixing, high dissolved oxygen concentrations (e.g. 60–80% saturation) could be demonstrated under conditions of oxygen limitation, i.e. high apparent 'critical' dissolved oxygen concentrations were observed. Such results could only be explained by assuming that the rate-limiting step in oxygen transfer was at the cell surface, i.e. either the liquid film resistance at the cell surface or the clump resistance (Bartholemew, W. H. *et al.*, 1950). Similar results were reported by Phillips, D. H. and Johnson, M. J. (1961) for *Penicillium chrysogenum* and *Aspergillus niger* cultures. Steel and Maxon also showed that, in such oxygen-limited cultures, the dissolved oxygen concentration increased with increasing impeller rotational speed or impeller diameter. This effect was attributed to the fact that increasing either parameter increased the oxygen solution rate (the rate of oxygen transfer from the gas phase into the culture fluid) to a proportionally greater extent than the rate of transfer from the fluid into the microbial cell, i.e. the primary effect was on the resistances at the gas/liquid interface. This situation is in direct contrast to that obtained with non-viscous Newtonian systems, where it is generally accepted that the resistances at the gas/liquid interface are rate-limiting. With non-viscous Newtonian systems, a low dissolved oxygen concentration (below the critical value) is taken as being indicative of oxygen-limiting conditions, whilst a high level is assumed to imply that the oxygen demand of the micro-organism is being satisfied. Steel and Maxon's studies indicate the danger of applying such criteria to viscous non-Newtonian systems; caution is obviously required in interpreting the effects of different dissolved oxygen concentrations, since a high concentration clearly does not necessarily mean that conditions are non-limiting with respect to oxygen transfer.

In an attempt to improve performance, Steel, R. and Maxon, W. D. (1966b) designed a multiple rod impeller which was considered to result in a more uniform distribution of shear rates throughout the fermenter. They found that the power consumed by the new design of impeller was substantially unaffected by the fluid viscosity whilst, under certain conditions, a greater oxygen transfer rate could be obtained than with conventional turbine impellers for the same power consumption.

Blakebrough, N. and Sambamurthy, K. (1966) have examined the relationships between power consumption, mixing time, impeller dimensions and $K_L a$ in laboratory fermenters, using both aerated and unaerated sulphite solution and sulphite solution containing 1·6% *w/v* paper pulp (to simulate the rheological properties of culture fluids of filamentous micro-organisms). They used two series of impellers in their experiments; in one series the overall diameter was varied, the blade dimensions being fixed whilst in the other series overall diameter was maintained constant and blade dimensions were varied. In aerated sulphite solutions, Blakebrough and Sambamurthy found that they were able to correlate $K_L a$ with power consumption irrespective of impeller dimensions, i.e. irrespective of the manner in which the power was transmitted. In the presence of paper pulp, however, impeller dimensions had a profound effect upon the volumetric transfer coefficient. They also found that in aerated and unaerated solutions, both in the presence and absence of paper pulp, mixing times could be related to a momentum factor, representing liquid flow from the impeller.

$$M = ND \times NWL(D-W) \tag{4.39}$$

where M = Momentum factor
W = Width of impeller blade
L = Length (height) of impeller blade

In the three-phase system (i.e. in the presence of paper pulp), Blakebrough and Sambamurthy were able to establish the following empirical correlation for K_La.

$$K_La = 152t_m{}^{0.203}\left(\frac{P}{V}\right)^{1.79} M^{-1.05} N^{-0.046} \tag{4.40}$$

The above correlation is of interest in that it includes impeller geometry. However, it is of limited practical value, since it does not include any fluid properties; indeed, Blakebrough and Sambamurthy only employed a single non-Newtonian fluid which was not rheologically characterized. Furthermore, the relationship is restricted to laboratory scale fermenters.

Other correlations for K_La. As mentioned previously, correlations between K_La and operating variables need not contain a term for power consumption, since this is determined by other factors. However, to be complete, such correlations should include all relevant fluid properties. Some correlations, which exclude power consumption and relate to non-Newtonian fluids, are briefly described here.

In 1961, Solomons, G. L. and Weston, G. O., using suspensions of the mould *Aspergillus*, which exhibited Bingham plastic behaviour, established an empirical relationship between oxygen transfer rate and both impeller shaft speed and mycelial concentration (or viscosity, since mycelial concentration and viscosity were directly proportional within the limits of the experiment); the oxygen transfer rate could thus be confidently predicted at any given impeller shaft speed and concentration of mycelium. This relationship was, of course, restricted to the organism and growth conditions in question and to the particular laboratory fermenter design employed.

Perez, J. F. and Sandall, O. C. (1974) studied gas absorption in conventionally designed laboratory scale fermenters in the presence of solutions of Carbopol (carboxy polymethylene): these solutions were pseudoplastic, obeying the power law. Perez and Sandall used carbon dioxide for their absorption studies in view of its relatively high solubility compared with oxygen and hence the higher absorption rates attainable. Gas absorption rates were determined at different impeller rotational speeds in the presence of solutions of Carbopol at three different concentrations. Perez and Sandall fitted their experimental results to a relationship previously derived by dimensional analysis by Sideman, S. *et al.* (1966) and obtained the following expression:

$$\frac{K_LaD^2}{D_L} = c\cdot\left(\frac{\rho ND^2}{\bar{\mu}_a}\right)^{1.11}\left(\frac{\bar{\mu}_a}{\rho D_L}\right)^{0.5}\left(\frac{DV_s}{\sigma}\right)^{0.447}\left(\frac{\mu_g}{\bar{\mu}_a}\right)^{0.694} \tag{4.41}$$

where μ_g = Gas viscosity

The average apparent viscosity ($\bar{\mu}_a$) was calculated using the relationship of Calderbank, P. H. and Moo-Young, M. B., 1959 (equation (4.36)). The groups in the above equation are a modified Sherwood Number, Reynolds Number, Schmidt Number, a gas flow group (which is not dimensionally sound) and the ratio gas viscosity/liquid viscosity. The work of Perez and Sandall is open to a number of criticisms. The principal criticism is that, although Carbopol solutions of three different concentrations were employed in their experiments, the data obtained from the two higher concentrations were excluded from the correlation due to excessive foaming; the correlation was thus based upon data obtained from a single Carbopol solution (0·25%) and water. Furthermore, only a single gas flow rate was employed throughout the experiments. In view of these limitations, their correlation cannot be regarded to be of general applicability. In addition, in view of the foaming problems encountered, even the 0·25% Carbopol solution may have contained traces of surface active material, which could possibly have affected their results.

A similar recent, but more comprehensive investigation, has been undertaken by Yagi, H. and Yoshida, F. (1975). Using a laboratory-scale fermenter of conventional design, with a single impeller and a capacity of approximately 12 litres, they investigated the effects of impeller rotational speed and air flow rate upon K_La (oxygen desorption) in the presence of different viscous liquids; these included aqueous solutions of glycerol and millet jelly (acid hydrolysate of starch) at different concentrations, both of which are Newtonian in character, and also aqueous solutions of sodium polyacrylate and sodium carboxymethylcellulose at different concentrations, which are non-Newtonian, pseudoplastic liquids obeying the power law. In the case of individual Newtonian fluids, Yagi and Yoshida found that the effect of impeller rotational speed and air flow rate upon K_La was consistent with the following equation:

$$K_La = k' \, N^{2\cdot2} \, V_s^{0\cdot28} \tag{4.42}$$

This equation is in approximate agreement with results obtained by other workers for sodium sulphite solutions (Cooper, C. M. *et al.*, 1944) and water (Yoshida, F. *et al.*, 1960). Yagi and Yoshida fitted their experimental data for Newtonian fluids to an expression obtained by dimensional analysis, similar to that of Sideman, S. *et al.* (1966) and obtained the following relationship:

$$\frac{K_LaD^2}{D_L} = c\cdot\left(\frac{\rho ND^2}{\bar{\mu}_a}\right)^{1\cdot5}\left(\frac{DN^2}{g}\right)^{0\cdot19}\left(\frac{\bar{\mu}_a}{\rho D_L}\right)^{0\cdot5}\left(\frac{\bar{\mu}_aV_s}{\sigma}\right)^{0\cdot6}\left(\frac{ND}{V_s}\right)^{0\cdot32} \tag{4.43}$$

The two last dimensionless groups in the above equation are a gas flow number after Johnson, D. L. *et al.* (1957) and the Aeration Number of Oyama, Y. and Endoh, K. (1955) to which reference has been made in an earlier section of this chapter. The exponent of 0·5 on the Schmidt Number was considered to be reasonable if the penetration or surface renewal theories for mass transfer across a gas/liquid interface hold (Higbie, R., 1935; Danckwerts, P. V., 1951). In the case of the non-Newtonian liquids employed, however, Yagi and Yoshida found that their data could not be fitted to this equation (the viscosity term used for these liquids was the average apparent viscosity after Metzner, A. B. and Otto, R. E., 1957). A good correlation could only be obtained if the above equation was modified to take account of the viscoelastic properties of the fluids in question. Yagi and Yoshida introduced a new dimensionless term, the Deborah Number, which contains the characteristic material time after Prest, W. M. *et al.* (1970), thus:

$$N_{De} = \lambda N \tag{4.44}$$

where N_{De} = Deborah Number
λ = Characteristic material time

The characteristic material time was defined by Prest *et al.* as the reciprocal of the shear rate at which the ratio of apparent viscosity to the zero-shear viscosity has a value of 0·67. Equation (4.43) was modified, in an empirical manner, by Yagi and Yoshida to yield the following expression:

$$\frac{K_L a D^2}{D_L} = c.\left(\frac{PND^2}{\bar{\mu}_a}\right)^{1\cdot 5}\left(\frac{DN^2}{g}\right)^{0\cdot 19}\left(\frac{\bar{\mu}_a}{pD_L}\right)^{0\cdot 5}\left(\frac{\bar{\mu}_a V_s}{\sigma}\right)^{0\cdot 6}\left(\frac{ND}{V_s}\right)^{0\cdot 32}$$
$$[1+2(\lambda N)^{0\cdot 5}]^{-0\cdot 67} \qquad (4.45)$$

The above equation correlated all their data for Newtonian and non-Newtonian fluids in a satisfactory manner. The last term of this expression reduces to unity in the case of Newtonian fluids. Although the experimental work of Yagi and Yoshida is restricted in the sense that the principal variables were impeller rotational speed, air flow rate and liquid viscosity (other parameters did not vary appreciably in their experiments) and it was not possible, of course, to vary each dimensionless group independently in the classical manner, their final correlation is nevertheless a comprehensive one and represents a valuable contribution to our knowledge in this field.

4.5 CONCLUSIONS

Over the last 20 years, developments have taken place which have considerably increased our understanding of aeration-agitation phenomena in the presence of viscous non-Newtonian fluids. Some of the more important developments have been described in this chapter. Much however remains to be done; this is still a relatively unexplored field. Perhaps the single greatest gap in our knowledge is the lack of data and correlations with actual mould and streptomycete culture fluids; much of the excellent work which has been performed has employed polymer solutions (e.g. the work of Metzner, A. B. and Otto, R. E., 1957 and Yagi, H. and Yoshida, F., 1975) and it is only assumed that the results of such work apply equally to mould and streptomycete culture fluids. It is only the latter fluids which, of course, are of interest to fermentation technologists.

In the author's opinion, further work in the following areas would be most rewarding:

(i) The technical difficulties involved in obtaining meaningful viscometric data from heterogeneous suspensions of filamentous micro-organisms.
(ii) Rheological information on mould and streptomycete culture fluids and the relationships between rheological properties and morphology in submerged culture.
(iii) Establishment of correlations for aeration efficiency ($K_L a$) and power consumption in the presence of mould and streptomycete culture fluids and, in particular, attempted confirmation of relationships previously established for polymer solutions.
(iv) Design of new, improved aeration-agitation systems, promoting

high oxygen transfer rates at low power consumption in the presence of viscous non-Newtonian fluids.

There would appear to be plenty of scope for further research by fermentation technologists and biochemical engineers, the results of which would be both interesting academically and of immense benefit to the fermentation and allied industries.

REFERENCES

Astarita, G. and Apuzzo, G. (1965), *A. I. Ch. E. Journal*, **11**, 815–820.

Banks, G. T., Mantle, P. G. and Szczyrbak, C. A. (1974), *J. Gen. Mic.*, **82**, 345–361.

Bartholemew, W. H., Karow, E. O., Sfat, M. R. and Wilhelm, R. H. (1950), *Ind. Eng. Chem.*, **42**, 1801–1809.

Bartholomew, W. H. (1960), *J. App. Microbiol.*, **2**, 289–300.

Blakebrough, N. and Sambamurthy, K. (1966), *Biotech. & Bioeng.*, **8**, 25–42.

Bongenaar, J. J. T. M., Kossen, N. W. F., Metz, B. and Meijboom, F. W. (1973), *Bioeng. & Biotech.*, **15**, 201–206.

Brierley, M. R. and Steel, R. (1959), *J. App. Microbiol.*, **7**, 57–61.

Calderbank, P. H. (1958), *Trans. Instn Chem. Engrs*, **36**, 443–463.

Calderbank, P. H. and Moo-Young, M. B. (1959), *Trans. Instn Chem. Engrs*, **37**, 26–33.

Calderbank, P. H. and Moo-Young, M. B. (1961), *Trans. Instn Chem. Engrs*, **39**, 338–347.

Casson, N. (1959), *Rheology of Disperse Systems*, Pergamon, 84.

Charm, S. E. (1963), *Ind. Eng. Chem. Process Des. Dev.*, **2**, 62.

Cheng, D. C. H. and Le Grys, G. A. (1975), *2nd Ann. Res. Meeting, Inst. Chem. Engnrs*, Bradford, March 1975.

Cooper, C. M., Fernstrom, G. A. and Miller, S. A. (1944), *Ind. Eng. Chem.*, **36**, 504–509.

Danckwerts, P. V. (1951), *Ind. Eng. Chem.*, **43**, 1460.

Deindoerfer, F. H. and Gaden, E. L. (1955), *J. App. Microbiol.*, **3**, 253–257.

Deindoerfer, F. H. and West, J. M. (1960), *J. Microbiol. Biochem. Technol. & Eng.*, **2**, 165–175.

Deindoerfer, F. H. and West, J. M. (1960), *App. Microbiol.*, **2**, 265–273.

Einstein, A. (1906), *Ann. Physik.*, **19**, 289.

Einstein, A. (1911), *Ann. Physik.*, **34**, 591.

Eirich, F. *et al.* (1936), *Kolloid Z.*, **74**, 276.

Finn, R. K. (1954), *Bact. Rev.*, **18**, 254–274.

Fondy, P. L. and Bates, R. L. (1961), Paper presented at A. I. Ch. E.–C.I.C. Meeting, Cleveland, U.S.A. May 1961.

Godleski, E. S. and Smith, J. C. (1962), *A. I. Ch. E. Journal*, **8**, 617–620.

Higbie, R. (1935), *Trans. A. I. Ch. E.*, **31**, 365.

Holland, F. A. and Chapman, F. S. (1966), *Liquid Mixing & Processing in Stirred Tanks*, 11–13, Van Nostrand Reinhold.

Johnson, D. L., Saito, H., Polejes, J. D. and Hougen, O. A. (1957), *A. I. Ch. E. Journal*, **3**, 411.

Johnstone, R. E. and Thring, M. W. (1957), *Pilot Plant Models & Scale-Up Methods*, 101, McGraw-Hill.

Karow, E. O., Bartholemew, W. H. and Sfat, M. R. (1953), *Agr. Food. Chem.*, **1**, 302–306.
Karow, E. O., Bartholemew, W. H. and Sfat, M. R. (1953), *Agr. Food Chem.*, **1**, 302–306.
Leduy, A., Marsan, A. A. and Coupal, B. (1974), *Biotech. Bioeng.*, **16**, 61–76.
Magnusson, K. (1952), *Iva* (Sweden), **23**, 86.
Maxon, W. D. (1959), *J. Biochem. Microbiol. Tech. & Eng.*, **1**, 311.
Metzner, A. B. and Otto, R. E. (1957), *A. I. Ch. E. Journal*, **3**, 3–10.
Metzner, A. B. and Taylor, J. S. (1960), *A. I. Ch. E. Journal*, **6**, 109–114.
Metzner, A. B., Feehs, R. H., Ramos, H. L., Otto, R. E. and Toothill, J. D., *A. I. Ch. E. Journal*, **7**, 3–9.
Michel, B. J. and Miller, S. A. (1962), *A. I. Ch. E. Journal*, **8**, 262–266.
Nishikawa, F. (1965), M.S. Thesis, University of Pennsylvania, Philadelphia, Pennsylvania, U.S.A.
Otto, R. E. (1957), Ph.D. Thesis, University of Delaware, Newark, U.S.A.
Oyamah, Y. and Endoh, K. (1955), *Chem. Eng.* (Japan), **10**, 2.
Perez, J. F. and Sandall, O. C. (1974), *A. I. Ch. E. Journal*, **20**, 770–775.
Phillips, D. H. and Johnson, M. J. (1961), *J. Biochem. Microbiol. Tech. Eng.*, **3**, 277–309.
Prest, W. M. and Porter, R. S. (1970), *J. App. Polymer. Sci.*, **14**, 2697–2706.
Richards, J. W. (1961), *Progr. Indust. Microbiol.*, **3**, 141–172.
Roels, J. A., Van Den Berg, J. and Voncken, R. M. (1974), *Biotech. Bioeng.* **16**, 181–208.
Rushton, J. H., Costich, E. W. and Everett, H. J. (1950), *Chem. Eng. Progr.*, **46**, 395–404.
Rushton, J. H., Costich, E. W. and Everett, H. J. (1950), *Chem. Eng. Progr.*, **46**, 467.
Rushton, J. H. (1951), *Chem. Eng. Progr.*, **47**, 485.
Sideman, S., Hortacsu, H. and Fulton, J. W. (1966), *Ind. Eng. Chem.*, **58**, 32–47.
Solomons, G. L. and Weston, G. O. (1961), *J. Biochem. Microbiol. Tech. Eng.*, **3**, 1–6.
Steel, R. and Maxon, W. D. (1962), *Biotech. Bioeng.*, **4**, 231–240.
Steel, R. and Maxon, W. D. (1966a), *Biotech. Bioeng.*, **8**, 97–108.
Steel, R. and Maxon, W. D. (1966b), *Biotech. Bioeng.*, **8**, 109–115.
Taguchi, H. and Miyamoto, S. (1966), *Biotech. Bioeng.*, **8**, 43–54.
Taguchi, H. and Kimura, T. (1970), *J. Ferment. Technol.*, **48**, 117.
Taylor, J. S. (1955) M.Ch.E. Thesis, University of Delaware, Newark, U.S.A.
Tuffile, C. M. and Pinho, F. (1970), *Biotech. Bioeng.*, **12**, 849–871.
Vand, V. (1948), *J. Phys. & Colloid Chem.*, **52**, 277.
Van Zanten, D. (1970), Internal Rept, Technological University of Delft, Netherlands.
Wang, D. I. C. and Humphrey, A. F. (1968), *Progr. Ind. Microbiol.*, **8**, 1–34.
Yagi, H. and Yoshida, F. (1975), *Ind. Eng. Chem. Process. Des. Dev.*, **14**, 488–493.
Yoshida, F. and Miura, Y. (1963), *Ind. Eng. Chem. Process. Des. Dev.*, **2**, 263.
Yoshida, F., Ikeda, A., Imakawa, S. and Miura, Y. (1968), *Ind. Eng. Chem.*, **52**, 435–438.

Chapter 5

Enzymic Alterations of Penicillins and Cephalosporins

Dr M. O. MOSS, Department of Microbiology, University of Surrey, Guildford, Surrey

5.1 INTRODUCTION

Since the first observations of Alexander Fleming in 1928 a large number of people, from an increasingly wide range of disciplines, have become involved in the study of the penicillins and related compounds. Something of the romance of the early studies of the penicillins is contained in the book *Mushrooms and Toadstools* by the late John Ramsbottom, who did so much to bring the study of fungi to a wider audience than the majority of scientists can command (Ramsbottom, 1953). The closely related group of antibiotics, the cephalosporins, have had an equally exciting history since the isolation of an antibiotic-producing strain of *Cephalosporium* by Brontzu in 1948 (Abraham and Loder, 1972). The penicillins have not only given rise to exciting possibilities in chemotherapy but have proved to be invaluable tools in such studies as the elucidation of the complex structure of the bacterial envelope (Parke, 1966).

5.1.1 The Penicillin Molecule

The penicillins are a family of compounds, each member of which differs by the nature of the acyl group attached to the 6-aminopenicillanic acid nucleus (6-APA) (Figure 5.1).

Despite the relatively small size of the penicillin molecule, the molecular weight of the sodium salt of benzylpenicillin is only 356: it contains a diverse range of functional groups making the chemistry of these compounds very complex and offering a wide range of possibilities for enzymic alterations (Figure 5.2).

The penicillins are produced by strains from a number of genera of micro-organisms other than *Penicillium*, including *Aspergillus*, *Trichophyton*, *Epidermophyton*, *Cephalosporium*, *Emericellopsis* and *Paecilomyces*. It has been established that the production of a tripeptide, α-aminoadipyl L-cysteinyl-L-valine is followed by cyclization to α-aminoadipyl 6-APA, in which the asymmetric centre of L-valine is changed to the D- configuration: however, the exact mechanism of this reductive cyclization is not understood. Further reactions then include transacylation to give the family of natural penicillins, and possibly deacylation to give free 6-APA.

R·NH, S, CH_3, CH_3, N, O, COOH

R	
C_6H_5–$CH_2 \cdot CO \cdot$	Penicillin G (benzylpenicillin)
$CH_3 \cdot CH_2 \cdot CH{=}CH \cdot CH_2 \cdot CO \cdot$	Penicillin F
$CH_3 \cdot (CH_2)_6 \cdot CO \cdot$	Penicillin K
HO–C_6H_4–$CH_2 \cdot CO$	Penicillin X

FIG. 5.1a Representatives of the natural penicillins which may be isolated from cultures of moulds growing on 'natural' substrates.

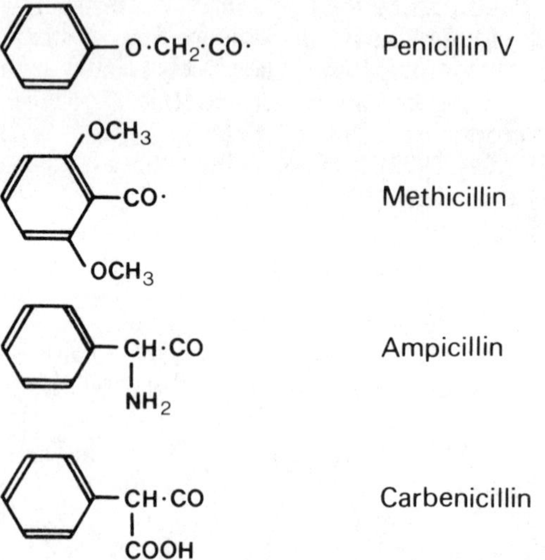

FIG. 5.1b Representatives of the semi-synthetic penicillins in which the side chain is chemically synthesized and added to 6-APA by a chemical process.

5.1.2 The Cephalosporin Molecule

The cephalosporins also form a family of compounds in which the amino group of 7-aminocephalosporanic acid (7-ACA) may be acylated with a range of carboxylic acids (Figure 5.3).

Although containing the beta lactam ring in common with the penicillins it is fused to a dihydrothiazine ring in the cephalosporins. The natural

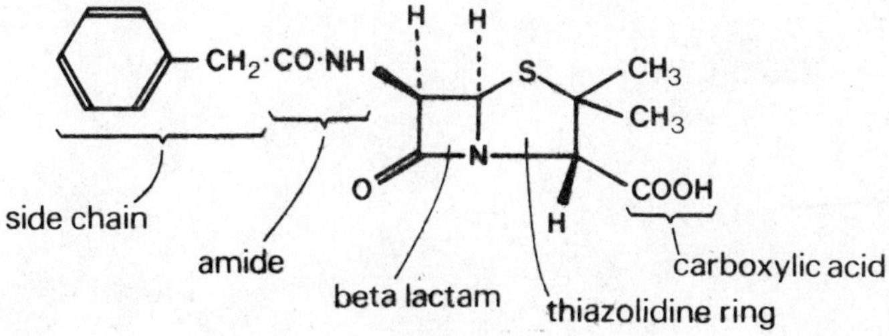

FIG. 5.2 N-phenylacetyl 6-aminopenicillanic acid (benzylpenicillin).

cephalosporins also contain a further functional group in the reactive allylic acetate side chain. Cephalosporin production is apparently very much more restricted in nature than is the production of penicillins. Amongst eucaryotes only a strain of *Cephalosporium* is known to produce cephalosporin C, the only naturally occurring member of the family produced by eucaryotes, although at least three related compounds have now been isolated from strains of the procaryotic genus *Streptomyces* (Higgens and Kastner, 1971; Nagarajan *et al.*, 1971). Two of these derivatives have an extra function at position 7 of the cephalosporanic acid nucleus (Figure 5.4).

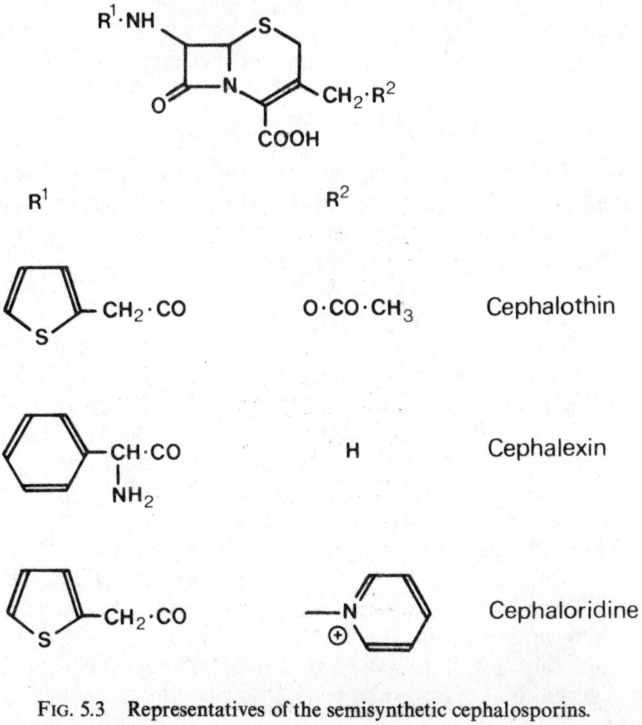

FIG. 5.3 Representatives of the semisynthetic cephalosporins.

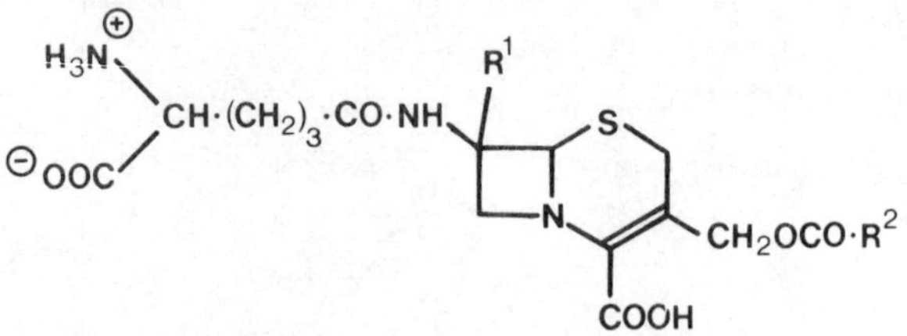

$R^1 = H$ $R^2 = CH_3$ Cephalosporin C

$R^1 = OCH_3$ $R^2 = CH_3$ Antibiotic from *Streptomyces lipmanii*

$R^1 = H$ $R^2 = NH_2$
$R^1 = OCH_3$ $R^2 = NH_2$ } Antibiotics from *Streptomyces clavuligerus*

FIG. 5.4 Natural cephalosporins.

5.1.3 Interconversion Between Penicillins and Cephalosporins

Because of the ready availability of the penicillin group of antibiotics, and the useful extension of properties offered by the cephalosporins, conversion from one series to the other would be a highly desirable process. Although the strain of *Cephalosporium* which produces cephalosporin C also produces a penicillin with the same alpha amino adipyl side chain (penicillin N), no microbiological or enzymic process for converting one to the other is known. However, a number of elegant chemical methods have been worked out for the synthesis of cephalosporins from penicillins (Cooper and Spry, 1972).

5.2 BETA LACTAMASES

The beta lactam function in both the penicillins and the cephalosporins is very sensitive to alkali, and in many of the penicillins to acid as well, so it is not perhaps surprising that this reactive ring can be hydrolysed by the enzymic activity of a large number of micro-organisms.

5.2.1 Penicillin Beta Lactamases (Penicillinases)

Penicillin proved to be very useful in the control of *Staphylococcus aureus* in the early days of penicillin therapy but, since the late 1940s, a rapid emergence of resistance occurred, especially in the hospital environment (Rolinson, 1971). This initial resistance of strains of *Staphylococcus aureus* was due to inactivation of penicillin by the enzyme penicillinase which opens the beta lactam ring to give the inactive penicilloic acid (Figure 5.5).

Penicillinases are now known to be produced by a number of bacteria from both Gram-positive and Gram-negative genera and their function and evolution have been analysed in some detail (Pollock, 1971). The primary structures of the penicillinases from *Staphylococcus aureus* (Amber and Medway, 1969) and *Bacillus licheniformis* (Meadway, 1969) have been analysed completely and they have been shown to be relatively small proteins containing a single polypeptide chain with no disulphide bridges.

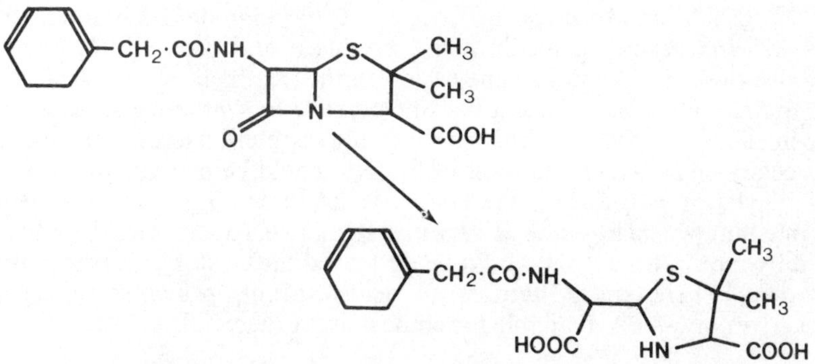

FIG. 5.5 Hydrolysis of benzylpenicillin by penicillinase.

Bacillus licheniformis produces not only a water soluble extracellular penicillinase, but, a proportion of the organisms' activity is retained bound to the plasma membrane of the bacterium. This membrane-bound material has been characterized, after isolation and purification, as a phospholipid containing protein with distinctly hydrophobic properties (Sawai and Lampen, 1974). The observations that the penicillinase activity of *Bacillus licheniformis* seemed to be coded by a single structural gene (Sherratt and Collins, 1973), and that the amino acid composition of both the membrane-bound phospholipid protein and the soluble exoenzyme are similar (Sawai and Lampen, 1974), indicate that one is a stage in the formation of the other.

An interesting frameshift mutation of the structural gene coding for the penicillinase of *B. licheniformis* produced an elongated penicillinase protein (Kelly and Brammar, 1973) in which the last three amino acid residues were replaced with seventeen new ones. Careful analysis of this mutant provided evidence that the wild type penicillinase is normally produced by proteolytic cleavage of a larger precursor molecule. The close structural relationship between the two forms of penicillinase produced by this organism has been further elucidated by Yamamato and Lampen (1975) who demonstrated that the hydrophobic membrane penicillinase differs from the hydrophylic exoenzyme by an additional phospholipopeptide chain of twenty-five amino acid residues terminating in a phosphatidyl serine. Bettinger and Lampen (1975) also obtained evidence for the existence of a protease sensitive, possibly partially folded, form of the enzyme, which was thought to be an intermediate in the process of penicillinase production, binding into membrane and subsequent secretion as the water-soluble form.

Unlike *Bacillus licheniformis*, penicillinase production by *Staphylococcus* is genetically controlled by extrachromosomal genes known as episomes (Novick and Richmond, 1965) and it is interesting to note that, although it is an inducible enzyme, the basal level is high when compared with other inducible enzyme systems. It has been suggested that, because staphylococci have a high intrinsic sensitivity to penicillins, and since enzyme induction is slow when compared with the rate of killing by penicillin, a high basal level of penicillinase will have some adaptive value (Novick, 1965). Once produced, the affinity of the enzyme for its substrate is very high as is the rate at which it catalyses the breakdown of penicillin. Thus the K_m value for benzylpenicillin is $2 \cdot 5 \times 10^{-6}$ M and turnover number 2×10^4 per minute (Richmond, 1963).

It was the natural penicillins first produced and used which, being so sensitive to penicillinase, became useless in the treatment of many infections due to *Staphylococcus*. Once it became possible to synthesize a wider range of penicillins it was found that, by introducing steric factors into the acyl side chain, penicillinase-resistant derivatives could be manufactured (Doyle and Nayler, 1964). Such a derivative is 2,6-dimethoxy phenyl penicillin (methicillin) which, however, is very unstable to acid and cannot be administered by mouth because it would be destroyed in the stomach before being absorbed. There are a number of acid-resistant, penicillinase-resistant derivatives of 6-APA available for clinical use (Figure 5.6).

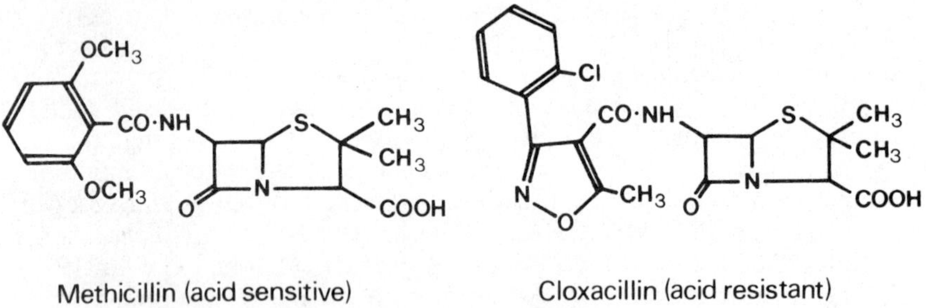

Methicillin (acid sensitive)

Cloxacillin (acid resistant)

FIG. 5.6 Examples of penicillinase resistant derivatives of 6-APA.

Although initially methicillin was active against all naturally occurring strains of *Staphylococcus aureus* tested, it was not long before resistant strains began to be encountered; and detailed studies of these have indicated that intrinsic resistance to methicillin, rather than ability to break down the antibiotic, was important (Sutherland and Rolinson, 1964). However, the relationship between penicillinase production and reaction to penicillinase-resistant derivatives of 6-APA is not a simple one. Lacey, Lewis and Grinstead (1973) have demonstrated that the antibiotic resistance of a hospital strain of *Staphylococcus aureus* was lost as a result of treatment with cloxacillin.

An interesting alternative to the use of penicillinase-resistant derivatives of 6-APA is indicated by the search for inhibitors of penicillinase which could then be used in conjunction with the penicillinase-sensitive penicillins which usually have a greater intrinsic activity. The isolation of such an inhibitor, named KA-107, from *Streptomyces gedamensis* has been described (Ohno *et al.*, 1973). A search for synthetic compounds, similar in structure to the penicillins, which would inhibit penicillinase has been unsuccessful (Baer and Mertes, 1973); and although the cephalosporins are competitive inhibitors of penicillinase, there is doubt about the value of using mixtures of penicillins and cephalosporins in therapy.

Although penicillinase production by bacteria is a nuisance from the point of view of chemotherapy, the enzyme does have some uses. Thus it is manufactured for the specific assay of penicillins and for the treatment of patients who are hypersensitive to penicillins and have been inadvertently given the antibiotic. For the assay of penicillins both biological and chemical methods depend on a difference in reactivity of the compound before and after opening the beta lactam ring. Some of the methods for estimating penicillins have been reviewed by Hamilton-Miller, Smith and Knox (1963). Penicillinase affords a highly specific method of obtaining a blank for such assays but the enzyme has also been used in the development of a penicillin electrode. Since the original concept of an enzyme electrode (Clark and Lyons, 1962) there have been several reports of such devices designed specifically for the assay of penicillins. They depend on the use of a hydrogen ion sensitive glass electrode incorporated into a compartment containing penicillinase which, although having access to the constituents of the solution to be assayed, is immobilized, either by a semi-permeable membrane, or by absorption on to a surface. Localized reaction of the enzyme with penicillin forms penicilloic acid and a measurable decrease in pH. Papariello, Mukherji and Shearer (1973) have described an electrode in which the penicillinase is immobilized in a layer of polyacrylamide gel moulded around, and in intimate contact with, a hydrogen ion glass electrode. Nilsson, Akerlund and Mosbach (1973) claim a linear response for their electrode system in the range 10^{-3} to 10^{-2} M penicillin and a reproducible response down to concentrations of 3×10^{-5} M. Cullen *et al.* (1974) describe an electrode system in which the penicillinase is adsorbed on to a fritted glass electrode fixed to the end of a flat surface pH sensitive glass electrode. Further development of such techniques offers the opportunity of the rapid estimation of penicillins which could be valuable, not only in the routine monitoring of a fermentation vessel, but in genetic and biochemical studies of penicillin activity and breakdown where large numbers of determinations often have to be made.

For the industrial production of penicillinase, *Bacillus cereus* is frequently used and is grown up in a well aerated medium containing corn steep liquor, yeast extract, phosphates and lard oil, for 18–24 hours at 30 °C. Enzyme production is induced by the addition of sterile penicillin solutions at intervals. After the required period of incubation the predominantly extracellular penicillinase is separated from the bacteria by centrifuging and subsequently precipitated from the clarified solution with acetone. The crude enzyme is redissolved in water, which must be pyrogen free if the enzyme is subsequently to be used for clinical purposes, and much of the impurities are removed by precipitating with ammonium sulphate. The penicillinase solution may then be dialysed against sodium chloride and vacuum dried. A valuable application of industrial preparations of *B. cereus* beta lactamase is in the estimation of antibiotics such as streptomycin in clinical material such as blood samples in the presence of penicillins (Waterworth, 1973).

Bacillus cereus in fact produces at least two beta lactamases, coded by two separate genes (Pollock and Fleming, 1969) and showing no immunological affinity. Detailed descriptions of the beta lactamases of both *B. cereus* and *B. licheniformis* have recently been published (Thatcher, 1975). Detailed studies of both the production and purification of penicillinases require sensitive assay techniques. A number of techniques depend on the fact that penicilloic acids, but not the corresponding penicillins, react rapidly with iodine. Thus a screening programme for penicillinase production might be carried out by growing the micro-organisms on nutrient agar plates which contain 1% soluble starch in the medium. After overnight growth the plates are flooded with a solution of benzylpenicillin and iodine in potassium iodide and the excess poured away. The presence of beta lactamase is indicated by a white halo spreading around the colony producing it (Richmond, 1975a). A critical review of the many methods now available for assaying beta lactamase activity has been given by Ross and O'Callaghan (1975).

Many Gram-negative bacteria produce penicillinases but, in contrast with the Gram-positive organisms so far discussed, the enzymes are almost invariably associated with the bacterial cell and their production is usually constitutive rather than inducible (Richmond and Sykes, 1973). At least some of the beta lactamases produced by Gram-negative bacteria are coded by part of the genome of resistance transfer factors (Jack and Richmond, 1970): thus Type IIIa beta lactamase has been found in a number of Gram-negative species, especially amongst the Enterobacteriaceae. The same enzyme has even been isolated from *Pseudomonas aeruginosa* (Sykes and Richmond, 1970) although most studies have been made on material from *Escherichia coli* (Richmond, 1975b). It has a broad substrate specificity including both penicillins and cephalosporins although showing a significantly higher affinity for the former.

5.2.2 Cephalosporin Beta Lactamases (Cephalosporinases)

Part of the initial interest in the cephalosporins stemmed from the observation that, those Gram-positive bacteria that owed their resistance to penicillins to their ability to produce penicillinase, are highly sensitive to the cephalosporins. Thus the penicillinases of *Staphylococcus aureus* and species of *Bacillus* have little effect on cephalosporins. *Bacillus cereus*, however, not only produces a beta lactamase which is a true penicillinase but also produces a second enzyme with a wider substrate specificity and activity against cephalosporins as well as penicillins. The situation amongst Gram-negative bacteria is much more complex; some strains producing beta lactamases specific for penicillins, others specific for cephalosporins and a third group which may hydrolyse either (Richmond and Sykes, 1973). The best studied cephalosporinases are those produced by species of *Enterobacter* (Ross, 1975).

The action of cephalosporinase on many of the cephalosporins is complex because the corresponding cephalosporoic acids produced by opening the beta lactam ring are usually unstable and fragment further (Hamilton-Miller, Newton and Abraham, 1970).

A strange fungal beta lactamase specific for the cephalosporins has been described from a species of *Cephalosporium* (Dennen, Allen and Carver, 1971).

5.3 PENICILLIN ACYLASES

The amide link joining the acyl side chain of the penicillins to 6-aminopenicillanic acid is more stable than the imide link of the beta lactam ring. Although it is now possible to effect the chemical cleavage of this amide link to liberate free 6-APA (Huber, Chauvette and Jackson, 1972), this reaction has been carried out on a large scale by the pharmaceutical industry using microbial enzymes (Figure 5.7). An extensive review and bibliography of penicillin acylase has been given by Vandamme and Voets (1974).

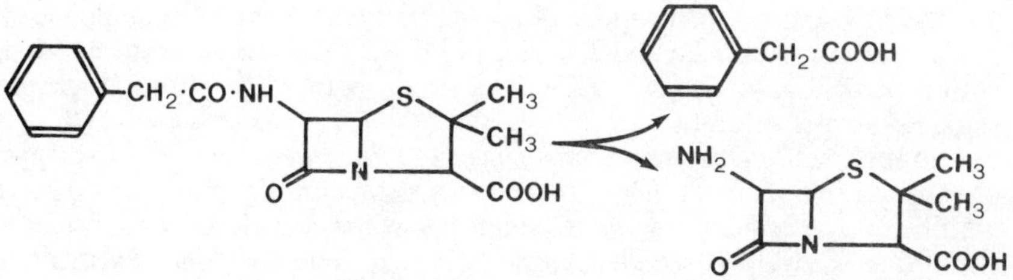

Fig. 5.7 Action of penicillin acylase on benzylpenicillin.

The hydrolysis of penicillin by a strain of *Penicillium chrysogenum* was first reported by Sakaguchi and Murao (1950) but without adequate characterization of the products. That 6-APA could indeed be produced in the growth medium of *P. chrysogenum* was confirmed by Batchelor, Doyle, Naler and Rolinson (1959) and the isolation of crystalline 6-APA subsequently described by Batchelor *et al.* (1961).

5.3.1 Occurrence in Nature

The production of an acylase enzyme (referred to as penicillin amidase in early reports) by bacteria was simultaneously reported by groups working in a number of different countries (Rolinson *et al.*, 1960; Claridge, Gourevitch and Lein, 1960; Huang *et al.*, 1960; Kaufmann and Bauer, 1960). Production of penicillin acylase is particularly widespread amongst Gram-negative bacteria, especially members of the Enterobacteriaceae. The list includes *Escherichia coli, Proteus rettgeri, Alcaligenes faecalis, Erwinea aroideae* and species of *Pseudomonas.* Amongst Gram-positive organisms activity has been recorded from *Micrococcus roseus* and *Bacillus megaterium*, the latter being particularly interesting in being extracellular (Chiang and Bennett, 1967). An extracellular acylase enzyme has also been described from the actinomycete *Streptomyces lavendulae* (Batchelor, Chain, Richards and Rolinson, 1961).

Many genera of fungi have now been associated with the ability to split one or more penicillins to liberate free 6-APA. These include *Penicillium, Aspergillus, Trichophyton, Cephalosporium, Emericellopsis, Fusarium* and *Botrytis.* Much of the early work on penicillin acylases has been reviewed by Hamilton-Miller (1966) and an extensive bibliography on the microorganisms associated with acylase activity has been given by Cole (1967).

5.3.2 Substrate Specificity of Penicillin Acylases

In general, fungal acylases hydrolyse phenoxymethylpenicillin faster than benzylpenicillin, whereas for the majority of bacteria the reverse is true. The exceptions amongst the bacteria include *Streptomyces lavendulae*, the extracellular enzyme of which has a substrate preference for phenoxymethylpenicillin (Batchelor *et al.*, 1961); which is also the preferred substrate for the acylase of *Erwinia aroideae* (Vandamme, Voets and Dhase, 1971). An interesting bacterial acylase, which has been partially purified from *Pseudomonas melanogenum*, showed a high specificity for α-aminobenzylpenicillin and did not hydrolyse either benzylpenicillin, p-aminobenzylpenicillin or phenoxymethylpenicillin (Okachi and Takishi, 1973). This enzyme is produced along with a penicillinase, the two activities being separated by chromatography using Sephadex columns.

Although both *Penicillium chrysogenum* and *Fusarium avenaceum* hydrolyse phenoxymethylpenicillin faster than benzylpenicillin, the particular pattern of substrate specificity varies considerably when a wide enough range of substrates is studied (Vanderhaeghe, Claesen, Vlietinck and Parmentier, 1968). For the series of n-alkyl penicillins *Fusarium avenaceum* shows an optimum rate of hydrolysis for the n-pentyl derivative, whereas in the case of *Penicillium chrysogenum* a steady increase in rate of hydrolysis is seen at least up to the n-undecyl derivative (Figure 5.8).

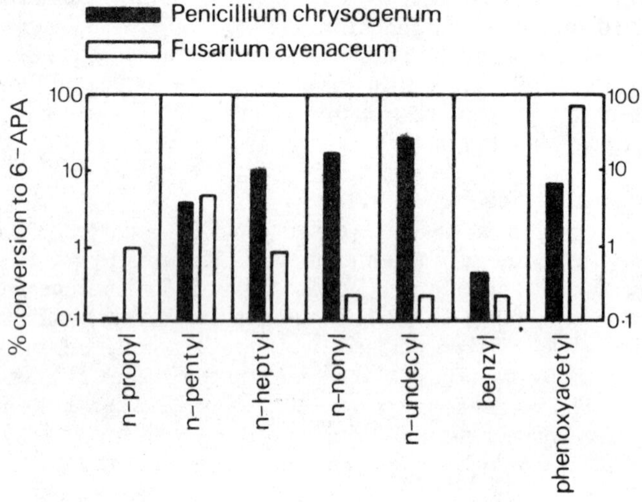

FIG. 5.8 Substrate specificity of the acylase enzymes of two fungi. The percentage conversion to 6-APA was estimated after 4 hr with reaction mixtures containing 0·25 mmole of penicillin + either 60 mg *F. avenaceum* extract (0·0364 units/mg) or 120 mg *P. chrysogenum* extract (0·0013 units/mg) in 20 ml water at 37 °C, pH 7·5. [*Vanderhaeghe*, et al., *1968*.

The enzyme from *Fusarium semitectum* has been purified to a considerable degree (Waldschmidt-Leitz and Bretzel, 1964). The mould, when grown in the presence of phenoxyacetic acid, showed a high acylase activity which was extracted from the dried mycelium with 0·2 N sodium acetate. It was then purified by a factor of × 300 to a nearly homogeneous protein which behaved as though it had a molecular weight of 65,000. The enzyme molecule contains two atoms of zinc, the removal of which leads to a loss of activity. The activity could be almost completely recovered by adding zinc sulphate solution and partially recovered by the addition of a number of other divalent metal salts.

An extensive study of the substrate specificity of the acylase enzyme of *Escherichia coli* has shown that, amongst penicillins, the rate of hydrolysis of benzylpenicillin is exceeded only by the rate of hydrolysis of p-hydroxybenzylpenicillin (Cole, 1969a). It has also been shown that the enzyme is not, in fact, a penicillin acylase specifically but that it will hydrolyse the acyl derivatives of a wide range of amino compounds, including ammonia itself (Cole, 1969b). Although a considerable degree of specificity exists when the nature of the acyl group is considered, the only element of specificity in the structure of the amino compound is that, if the amino group is attached to an asymmetric centre, it should have the L-configuration. 6-aminopenicillanic acid has just such a configuration at the carbon atom to which the amino group is attached and so falls in the range of possible substrates for this enzyme. Although the geometry of the acyl substituent cannot be the only factor involved in the specificity which the enzyme shows for the acyl group, it is interesting to note that, when the extended length, as measured from Van der Waals models, is plotted against efficiency of hydrolysis, a sharp optimum is observed (Figure 5.9).

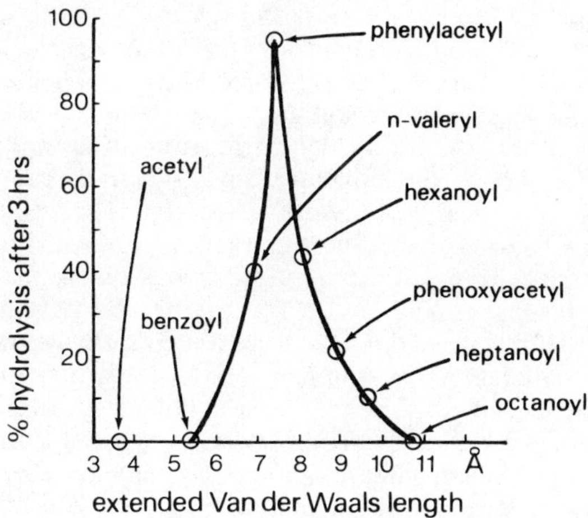

FIG. 5.9 The relationship between the Van der Waals length of the acyl group and the rate of hydrolysis of acyl derivatives of glycine by the acylase enzyme of *Escherichia coli*.

The reaction catalysed by the acylase of *E. coli* is reversible and the enzyme can be used for the synthesis of a number of acyl derivatives of 6-APA (Cole, 1969c). The optimum conditions for the synthetic reaction are different from those for the splitting reaction: thus, in the case of the enzyme of *E. coli*, the rate of synthesis of benzylpenicillin from 6-APA and phenylacetic acid is optimal at about pH 5·0, whereas the rate of hydrolysis of benzylpenicillin is optimal at about pH 8·5. The enzyme will act as an acyl transferase if derivatives such as amides, methyl esters or N-glycyl compounds are used, instead of the free acids, in the synthesis of penicillins. Indeed by using such derivatives of carboxylic acids the range of penicillins which can be synthesized is increased: thus, whereas phenoxymethylpenicillin is not produced from 6-APA and phenoxyacetic acid using the *E. coli* enzyme, it is produced from 6-APA and N-phenoxyacetylglycine.

The acylase enzyme of *Escherichia coli* has been purified and crystallized (Kutzbach and Rauenbusch, 1974). It has been characterized as having a molecular weight of 70,000 but in 1% sodium dodecyl sulphate the enzyme is partially and reversibly dissociated into units of molecular weight about 20,500. This purified enzyme showed the same range of substrate specificity as the crude cell-bound enzyme of earlier studies. Because the acylase enzyme is of considerable importance in the manufacture of 6-APA, itself a precursor of the wide range of semi-synthetic penicillins available for present-day chemotherapy, there have been considerable screening programmes for this enzyme in many laboratories throughout the world. Because of the range of substrates with which the enzyme will react a number of colorimetric assays have been developed to aid such screening programmes. The hydrolysis of phenylacetyl L-asparagine, coupled with the hydrolysis of the asparagine released using asparaginase, provides a sensitive assay for the acylase, the ammonia formed being determined by Nesslerization techniques (Bauer, Kaufmann and Ludwig, 1971). The incorporation of Phenylacetyl 4-nitro anilide into agar plates has given a useful qualitative method for the initial screening of micro-organisms (Walton, 1964). Kutzbach and Rauenbusch (1974) have used the increase in optical density at 405 nm following the hydrolysis of the water soluble sodium salt of 6-nitro-3-phenylacetamido-benzoic acid as an assay during the purification of *E. coli* acylase. Using penicillins themselves as the substrate, acylase activity can be assayed by methods designed to measure the 6-APA liberated. The well tried hydroxylamine assay can be used after solvent extraction of residual penicillin at low pH (Batchelor *et al.*, 1961). The method depends on the fact that the beta lactam ring of penicillins and 6-APA reacts with hydroxylamine at neutral pH to form a hydroxamic acid which forms a reddish-brown compound with Ferric ions. If the beta lactam ring has previously been opened, either with alkali or more specifically with penicillinase, it no longer reacts with hydroxylamine. Because 6-APA is so readily separated from most penicillins by paper chromatography, a biochromatographic assay may also be used to follow the hydrolysis of a penicillin by a preparation suspected of containing an acylase. Duplicate chromatograms are prepared on strips of paper. One of the chromatograms is sprayed, first with aqueous sodium bicarbonate, then with a solution of phenylacetyl chloride in acetone. Once the tapes are thoroughly dry they are carefully laid on a sheet of nutrient agar seeded with spores of *Bacillus subtilis*. The presence of 6-APA is demonstrated by the appearance of a new or increased zone of inhibition after spraying with phenylacetyl chloride which will have converted any 6-APA to benzylpenicillin. A critical review of these and other assay methods has been presented by Cole, Savidge and Vanderhaeghe (1975).

5.3.3 Industrial Production and Use of Penicillin Acylase

Although 6-APA can be produced by direct fermentation, the yield is inevitably very low and all the evidence from studies on the biosynthesis of the penicillins indicates that 6-APA is a side product rather than a precursor. The modern penicillin industry requires a continuous supply or pure 6-APA which is now manufactured by enzymic or chemical splitting of the readily available benzylpenicillin. Some of the factors in the development of commercial processes for the production of the raw material, 6-APA, are discussed by Carrington (1971). Using the acylase of *Escherichia coli*, which is usually cell-bound, it has been traditional to grow up the organism under conditions in which the enzyme is induced, add a concentrated solution of benzylpenicillin to a suspension of the washed cells under controlled conditions of pH and temperature, remove the bacteria and discard them. There are, however, advantages in solubilizing and purifying the enzyme to some extent and then rebinding it to an inert polymer. In this form the enzyme is usually more stable and can be reused several times in a batch process or incorporated into a continuous process using columns. The use of semi-purified immobilized enzymes rather than intact cells is likely to give a cleaner product. Technical details for the binding of *Escherichia coli* acylase to a cyanogen bromide-activated Sepharose are described by Savidge and Cole (1975). The final product is stored damp at 4 °C, under which conditions it is quite stable. These authors point out that the properties of an enzyme may be altered by binding, depending upon the nature of the inert carrier and the method used for attaching the enzyme to it.

The conditions in which *E. coli* is grown for the industrial production of acylase are a little different from the familiar bacteriological picture of this organism as a facultative anaerobe able to vigorously ferment sugars and grow optimally at 37 °C in the guts of animals. The enzyme is produced in best yield under mildly aerobic conditions (production is inhibited if aeration is too vigorous) using media free of fermentable carbohydrates and at a temperature controlled at 24 °C. Production of the enzyme is repressed by the presence of carbohydrates but readily induced by the presence of phenylacetic acid. The importance of the inducer is emphasized by a report of Cole and Sutherland (1966) that, of 148 strains of bacteria studied only one produced acylase in the absence, and ten in the presence, of phenyl acetic acid.

A typical pilot plant production might proceed as follows: 590 litres of a medium containing 13·3 kg of corn steep liquor, a valuable waste product from the manufacture of starch from maize, are inoculated with 3 litres of a vegetative seed of *E. coli*. After 8 hrs aerated growth at 24 °C, 600 ml portions of sterile approximately 25% aqueous ammonium phenylacetate are added at hourly intervals for a further 13 hrs. The bacteria are killed and harvested at 22–28 hrs and used directly as a crude enzyme preparation. A slurry of *E. coli* obtained in this way may then be added to a 6% solution of benzylpenicillin adjusted to pH 8·0 and the reaction mixture stirred at 37 °C. Because of the release of phenylacetic acid by the reaction, the pH must be maintained by the continuous addition of alkali; a check on the volume of alkali added also serving to monitor the progress of the reaction. When the reaction is completed, the bacteria are removed, the reaction mixture adjusted to pH 2, extracted with solvent to remove phenylacetic acid and any unchanged benzylpenicillin, and readjusted to pH 4·3, the isoelectric point of the amphoteric 6-APA, at which it may precipitate as a fine white crystalline material. It is not always necessary to isolate and purify the 6-APA. If a penicillin is readily obtained by acylation of 6-APA in aqueous solution, by for example an acyl chloride, and the resulting penicillin is readily extracted and separated from phenyl acetic acid, then the crude preparation of acylase may be added directly to the broth in which benzylpenicillin has been produced.

5.4 OTHER ENZYMES FOR WHICH PENICILLINS AND CEPHALOSPORINS MAY BE SUBSTRATES

The acetylated hydroxymethyl group in the natural antibiotic, cephalosporin C (Figure 5.4) is fairly sensitive to hydrolysis. Huber, Baltz and Caltrider (1968) demonstrated that traces of deacetyl cephalosporin C present in the growth media of *Cephalosporium* were probably due to non-enzymic hydrolysis; but enzymic deacetylation has been observed using an esterase from citrus fruits (Jeffery, Abraham and Newton, 1961) and from a number of micro-organisms (Demain, Walton, Newkirk and Miller, 1963). The esterase from citrus fruit is not specific and can be assayed using triacetin as substrate. It is found in highest concentrations in the outer layers of the peel of oranges, and details of its extraction and partial purification are given by Abraham and Fawcett (1975). A good source of the enzyme has been reported to be from strains of *Bacillus subtilis* grown in trypticase soy broth (Abbott and Fukuda, 1975). The cephalosporin acetylesterase occurs in both the culture supernatant and associated with the bacterial cells; and, although only about 25% of the total activity is liberated in the supernatant, it has been found to be more convenient to recover and purify the enzyme from the clarified broth. The purified enzyme behaves on a Sephadex column as though it had a molecular weight of about 190,000 (Abbott and Fukuda, 1975). Unlike the citrus fruit enzyme, the esterase from *B. subtilis* is remarkably stable and neutral solutions can be kept at room temperature for as long as three weeks with little loss of activity.

It is sometimes necessary to protect the carboxyl group of 6-APA during the synthesis of semisynthetic penicillins. If this is done by the formation of an ester the protecting group may be removed subsequently by hydrolysis under mild conditions using microbial enzymes. The esters of penicillins have been of interest in the past because, although they do not usually have anything like the *in vitro* activity of the free carboxylic acids, they are usually fat soluble and sometimes more stable than the corresponding penicillins. Despite their low *in vitro* activity, compounds such as the ethyl ester of benzylpenicillin (Figure 5.10) do have a high activity in protecting some,

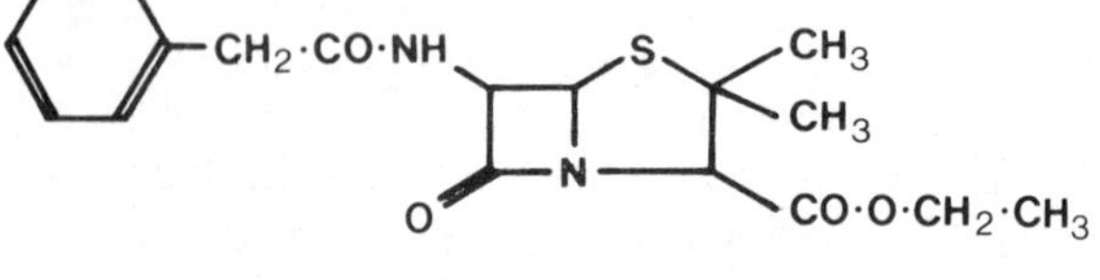

$CH_2 \cdot CO \cdot NH$ S CH_3 CH_3 N O $CO \cdot O \cdot CH_2 \cdot CH_3$

FIG. 5.10 Ethyl ester of benzylpenicillin.

but not all, animal species against infections such as those caused by *Streptococcus haemolyticus*. The specificity amongst different animal species has been shown to be due to the fact that the serum of animals such as the mouse and rat is able to hydrolyse the esters to the free, highly active, penicillin. The serum of animals such as the rabbit, dog, monkey and man does not contain such esterases and the penicillin remains in the body as the intact relatively inactive ester (Richardson, Walker, Miller and Hansen, 1945; Kirchner, McCormick, Cavallito and Miller, 1949).

The side chain attached by the amide link to either 6-APA or 7-ACA may be altered without being removed by enzymic activity. In the animal body aromatic functions in the side chain may be hydroxylated, presumably as a preliminary to the antibiotic being excreted. The D-α-amino adipyl side chain of cephalosporin C is oxidized by an enzyme produced by a strain of *Aspergillus flavus* (Fildes, Arnold and Gilbert, 1969). The final result of the oxidation is a cephalosporin with an α-ketoadipyl side chain.

During a study of the transpeptidase enzymes of strains of *Streptomyces* an extracellular enzyme was discovered from one strain which has been reported to fragment benzylpenicillin with the liberation of phenylacetylglycine (Frère *et al.*, 1975). It is suggested that a cleavage such as that indicated in Figure 5.11 is responsible for the formation of phenylacetylglycine,

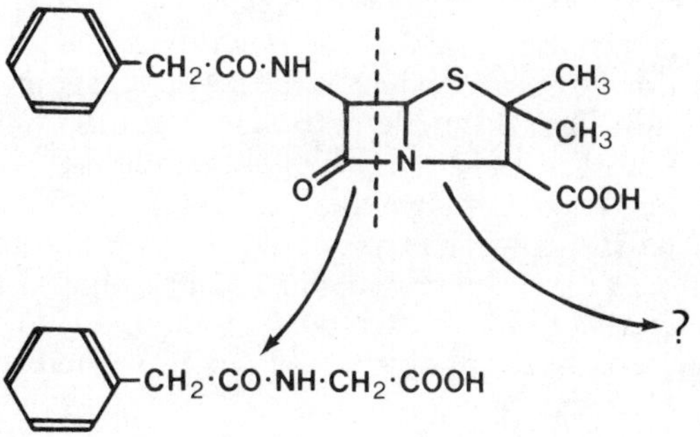

FIG. 5.11 Suggested fragmentation of benzylpenicillin by a strain of *Streptomyces*.

although the mechanism suggested is very tentative. A review and bibliography of the studies on the interaction of streptomycetes with penicillins, including a description of the beta lactamases produced by some strains of this group of procaryotes, is given by Johnson *et al.* (1975).

5.5 NON-ENZYMIC REACTIONS OF 6-APA IN CULTURE MEDIA

The chemistry of the penicillins and cephalosporins is complex, and some details of many of the diverse chemical reactions which the fused beta lactam-thiazolidine, or dihydrothiazine, structures can undergo are described by Doyle and Nayler (1964). Further details of the elegant rearrangements of which these structures are capable are given by Barton and Sammes (1971) and in various chapters of Flynn (1972). Some of the non-enzymic reactions which 6-APA, or related compounds, may undergo are relevant to the products which one might expect to find when studying the microbial alteration of these compounds.

5.5.1 Reaction with Carbon Dioxide

When 6-APA is added to solutions containing reducing sugars, derivatives some of it is decomposed to a compound which can no longer be converted

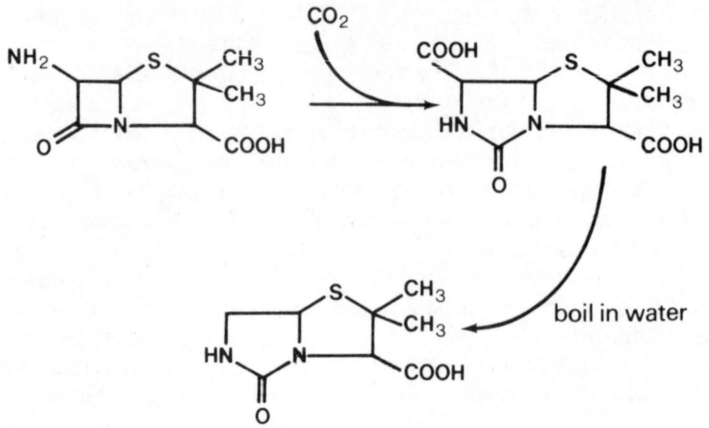

Fig. 5.12 The reaction of 6-APA with carbon dioxide.

into penicillins by acylation (Batchelor, Gazzard and Nayler, 1961). It is, in fact, the carbon dioxide released which is the active compound for it has been shown that, when this gas is bubbled through a solution of 6-APA at 37 °C and an initial pH of 7·0, it has a half-life of about 3 hrs and is converted into a compound no longer containing either the beta lactam ring or the primary amino group. The rearrangement proposed by these authors is shown in Figure 5.12 and was independently confirmed by Johnson and Hardcastle (1961). As would perhaps be expected this same compound has been isolated from natural fermentations in which 6-APA and carbon dioxide are produced (Ballio *et al.*, 1961).

5.5.2 Reactions with Sugars

When 6-APA is added to solutions containing reducing sugars, derivatives are formed in which the beta lactam ring, although still present, is more stable to penicillinase. The compounds formed, which are readily hydrolysed by acid to free 6-APA and sugar, were characterized as N-glycosyl derivatives (Moss and Cole, 1964), a typical example of which is shown in Figure 5.13.

Compounds, with properties similar to those formed using pure sugars and 6-APA, were identified in both the culture filtrates and the mycelium of *Penicillium chrysogenum* when grown under conditions in which it produced 6-APA. These glycosyl derivatives have little if any biological activity and, as they do not react with phenylacetyl chloride unless conditions have become sufficiently acid for liberation of free 6-APA to occur, they represent a bound form of 6-APA which may be missed during bioassays of media containing sugars. Although the formation of an N-glycosyl derivative is the primary reaction, Moss and Cole (1964) also reported a loss of beta lactam nucleus

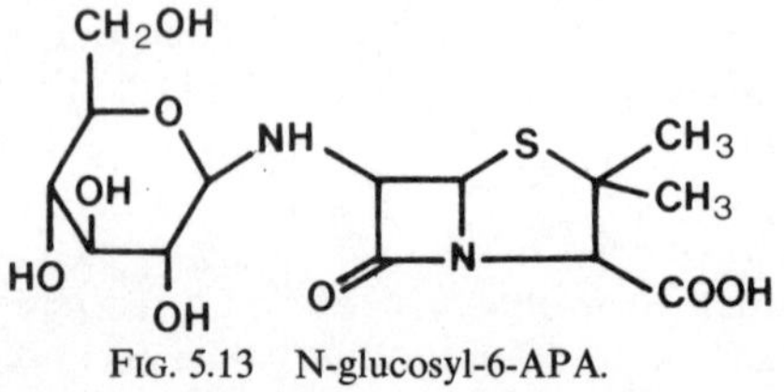

Fig. 5.13 N-glucosyl-6-APA.

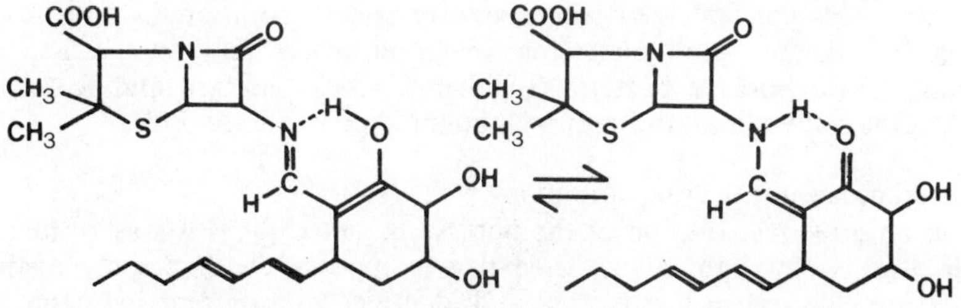

FIG. 5.14 Reaction product formed from 6-APA and frequentin.

when 6-APA is mixed with reducing sugars. This effect was particularly marked with mannose and represents a nonenzymic destruction of the beta lactam ring at neutral pH.

5.5.3 Reactions with other Fungal Metabolites

During a programme of work in which 6-APA was added to growing cultures of micro-organisms in order to identify strains which might alter the penicillin nucleus, a mould producing the metabolite frequentin was used. A new derivative was isolated which behaved chromatographically like a penicillin, but with little of the antibiotic activity of a true penicillin. Spectroscopic and chemical evidence indicated that this new compound was a Schiff's base of the type shown in Figure 5.14 (Moss, 1964).

Undoubtedly a number of fungi, under certain conditions of growth, will produce carbonyl compounds which could interact with 6-APA. Experiments with model compounds indicated that such derivatives could be sufficiently stable if hydrogen bonding and resonance of the type indicated for the frequentin derivative in Figure 5.14 were possible. Thus the compound of 6-APA with 2-hydroxynaphthaldehyde has been isolated and characterized by Wolfe, Godfrey and Perron (1963).

5.5.4 Polymerization

Concentrated aqueous solutions of 6-APA slowly deteriorate with the formation of fairly high molecular weight polymers by the addition of the amino group of one molecule to the beta lactam group of another (Grant, Clark and Alburn, 1962). Such polymers have no antibiotic activity against *Staphylococcus aureus* either before or after treatment with phenylacetyl

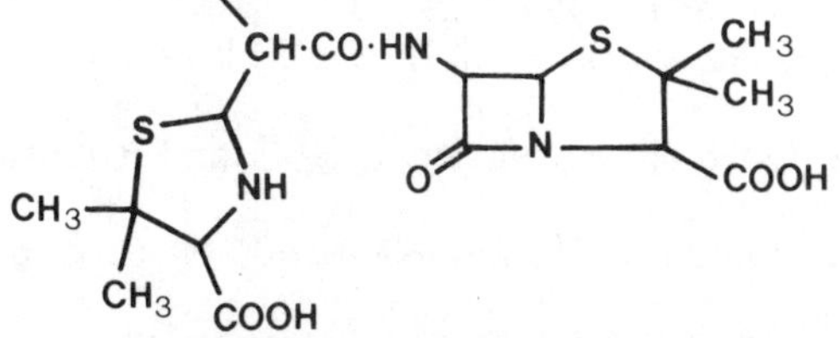

FIG. 5.15 Structure of the dimer of 6-APA.

chloride. Less concentrated solutions give rise to compounds with lower molecular weights and possessing some biological activity, particularly against Gram positive bacteria (Batchelor, Cole, Gazzard and Rolinson, 1962). One such compound may well be the dimer (Figure 5.15).

5.5.5 Photochemical Transformations

When an aqueous solution of the potassium salt of 6-APA was irradiated with ultra violet light at 15 °C, a new compound with activity against *Staphylococcus aureus* was formed and identified as aminomethyl penicillin (Godtfredsen, von Daehne and Vangedal, 1967). It was suggested that fission of the beta lactam ring gives rise to the formation of an aminoketene which subsequently reacts with a second molecule of 6-APA to form the new penicillin (Figure 5.16).

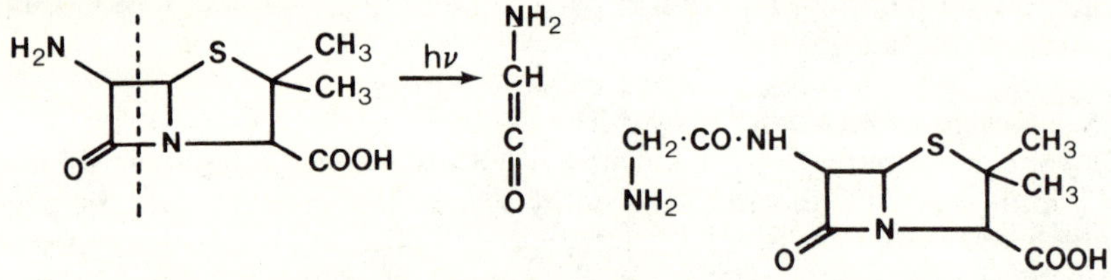

H_2N S CH_3 CH_3 N O COOH $\xrightarrow{h\nu}$ NH_2–CH=C=O

$CH_2 \cdot CO \cdot NH$ NH_2 S CH_3 CH_3 N O COOH

FIG. 5.16 Formation of aminomethylpenicillin.

REFERENCES

Abbott, J. B. and Fukuda, D. S. (1975), *Methods in Enzymology*, **43**, 731–734.

Abraham, E. P. and Fawcett, P. (1975), *Methods in Enzymology*, **43**, 728–731.

Abraham, E. P. and Loder, P. B. (1972), in *Cephalosporins and Penicillins, Chemistry and Biology*. Ed. by E. H. Flynn, Academic Press, N.Y.

Ambler, R. P. and Meadway, R. J. (1969), *Nature*, **222**, 24–26.

Baer, T. A. and Mertes, M. P. (1973), *J. med. Chem.*, **16**, 85–87.

Ballio, A., Chain, E. B., Dentice di Accadia, F., Mauri, M., Rauer, K., Schlesinger, M. J. and Schlesinger, S. (1961), *Nature*, **191**, 909–910.

Barton, D. H. R. and Sammes, P. G. (1971), *Proc. R. Soc. London B*, **179**, 345–355.

Batchelor, F. R., Chain, E. B., Hardy, T. L., Mansford, K. R. L. and Rolinson, G. N. (1961), *Proc. R. Soc. London B*, **154**, 498–508.

Batchelor, F. R., Chain, E. B., Richards, M. and Rolinson, G. N. (1961), *Proc. R. Soc. London B*, **154**, 522–531.

Batchelor, F. R., Cole, M., Gazzard, D. and Rolinson, G. N. (1962), *Nature*, **195**, 954–955.

Batchelor, F. R., Doyle, F. P., Nayler, J. H. C. and Rolinson, G. N. (1959), *Nature*, **183**, 257–258.

Batchelor, F. R., Gazzard, D. and Nayler, J. H. C. (1961), *Nature*, **191**, 910–911.

Bauer, K., Kaufmann, W. and Ludwig, St. A. (1971), *Hoppeseyler's Z. Physiol. Chem.*, **352**, 1723–1724.

Bettinger, G. E. and Lampen, J. O. (1975), *J. Bacteriol.*, **121**, 83–90.
Carrington, T. R. (1971), *Proc. R. Soc. London B*, **179**, 321–333.
Chiang, C. and Bennett, R. E. (1967), *J. Bacteriol.*, **93**, 302–308.
Claridge, C. A., Gourevitch, A. and Lein, J. (1960), *Nature*, **187**, 237–238.
Clark, L. C. and Lyons, C. (1962), *Ann. N.Y. Acad. Sci.*, **102**, 29–45.
Cole, M. (1964), *Nature*, **203**, 519–520.
Cole, M. (1967), *Process Biochem.*, **2**(4), 35–41.
Cole, M. (1969a), *Biochem. J.*, **115**, 733–739.
Cole, M. (1969b), *Biochem. J.*, **115**, 741–745.
Cole, M. (1969c), *Biochem. J.*, **115**, 747–756.
Cole, M., Savidge, T. and Vanderhaeghe, H. (1975), *Methods in Enzymology*, **43**, 698–705.
Cole, M. and Sutherland, R. (1966), *J. gen. Microbiol.*, **42**, 345–356.
Cooper, R. D. G. and Spry, D. O. (1972), in *Cephalosporins and Penicillins, Chemistry and Biology*. Ed. by E. H. Flynn, Academic Press, N.Y.
Cullen, L. F., Rusling, J. F., Schleifer, A. and Papariello, G. J. (1974), *Anal. Chem.*, **46**, 1955–1961.
Demain, A. L., Walton, R. B., Newkirk, J. F. and Miller, I. M. (1963), *Nature*, **199**, 909–910.
Dennen, D. W., Allen, C. C. and Carver, D. D. (1971), *J. Appl. Microbiol.*, **21**, 907–915.
Doyle, F. P. and Nayler, J. H. C. (1964), *Advances in Drug Research*, **1**, 1–69.
Fildes, R. A., Arnold, B. H. and Gilbert, D. A. (1969), Belgium Patent No. 736,934.
Flynn, E. H. (1972), *Cephalosporins and Penicillins, Chemistry and Biology*. Ed. by E. H. Flynn, Academic Press, N.Y.
Frère, J.-M., Ghuysen, J.-M., Degelaen, J., Loffet, A. and Perkins, H. R. (1975), *Nature*, **258**, 168–170.
Godtfredsen, W. O., von Daehne, W. and Vangedal, S. (1967), *Experientia*, **23**, 280–281.
Grant, N. H., Clark, D. E. and Alburn, H. E. (1962), *J. Amer. Chem. Soc.*, **84**, 876–877.
Hamilton-Miller, J. M. T. (1966), *Bacteriol. Rev.*, **30**, 761–771.
Hamilton-Miller, J. M. T., Newton, G. G. F. and Abraham, E. P. (1970), *Biochem. J.*, **116**, 371–384.
Hamilton-Miller, J. M. T., Smith, J. T. and Knox, R. (1963), *J. Pharm. Pharmacol.*, **15**, 81–91.
Higgens, C. E. and Kastner, R. E. (1971), *Int. J. Syst. Bacteriol.*, *21*, 326–331.
Huang, H. T., English, A. R., Seto, T. A., Shull, G. M. and Sobin, B. A. (1960), *J. Amer. Chem. Soc.*, **82**, 3790–3791.
Huber, F. M., Baltz, R. H. and Caltrider, P. G. (1968), *Appl. Microbiol.*, **16**, 1011–1014.
Huber, F. M., Chauvette, R. R. and Jackson, B. G. (1972), in *Cephalosporins and Penicillins, Chemistry and Biology*. Ed. by E. H. Flynn, Academic Press, N.Y.
Jack, G. W. and Richmond, M. H. (1970), *J. gen. Microbiol.*, **61**, 43–61.
Jeffery, J. D'A., Abraham, E. P. and Newton, G. G. F. (1961), *Biochem. J.*, **81**, 591–596.

Johnson, D. A. and Hardcastle, G. A. (1961), *J. Amer. Chem. Soc.*, **83**, 3534–3535.

Johnson, K., Duez, C., Frère, J.-M. and Ghuysen, J.-M. (1975), *Methods in Enzymology*, **43**, 687–698.

Kaufmann, W. and Bauer, K. (1960), *Naturwissenschafen*, **47**, 474–475.

Kelly, L. E., and Brammar, W. J. (1973), *J. Mol. Biol.*, **80**, 135–147.

Kirchner, F. K., McCormick, J. R., Cavallito, C. J. and Miller, L. C. (1949), *J. Org. Chem.*, **14**, 388–393.

Kutzbach, C. and Rauenbusch, E. (1974), *Hoppeseyler's Z. Physiol. Chem.*, **355**, 45–53.

Lacey, R. W., Lewis, E. and Grinstead, J. (1973), *J. med. Microbiol.*, **6**, 191–199.

Meadway, R. J. (1969), *Biochem. J.*, **115**, 12*P*–13*P*.

Moss, M. O. (1964), *Experientia*, **20**, 605–606.

Moss, M. O. and Cole, M. (1964), *Biochem. J.*, **92**, 643–648.

Nagarajan, R., Boeck, L. D., Gorman, M., Hamill, R. L., Higgens, C. E., Hoehn, M. M., Stark, W. M. and Whiney, J. G. (1971), *J. Amer. Chem. Soc.*, **93**, 2308–2310.

Nilsson, H., Akerlund, A-C. and Mosbach, K. (1973), *Biochim. Biophys. Acta*, **320**, 529–534.

Novick, R. P. (1965), *Annals N.Y. Acad. Sci.*, **128**, 165–182.

Novick, R. P. and Richmond, M. H. (1965), *J. Bacteriol.*, **90**, 467–480.

Ohno, H., Matsumae, A., Iwai, Y., Nakae, M., Omura, S. and Hata, T. (1973), *Antimicrob. Agents Chemother.*, **4**, 226–230.

Okachi, R. and Takishi, N. (1973), *Agri. Biol. Chem.*, **37**, 2797–2804.

Papariello, G. J., Mukherji, A. K., and Shearer, C. M. (1973), *Anal. Chem.*, **45**, 790–792.

Park, J. T. (1966), in 'Biochemical Studies of Antimicrobial Drugs', *16th Symposium of the Soc. gen. Microbiol*, Cambridge University Press.

Pollock, M. R. (1971), *Proc. R. Soc. London, Ser. B*, **179**, 385–401.

Pollock, M. R. and Fleming, J. (1969), *J. gen. Microbiol.*, **59**, 303–316.

Ramsbottom, J. (1953), *Mushrooms and Toadstools; a Study of the Activities of Fungi* (The New Naturalist Series), Wm. Collins.

Richardson, A. P., Walker, H. A., Miller, I. and Hansen, R. (1945), *Proc. Soc. Exptl. Biol. Med.*, **60**, 272–276.

Richmond, M. H. (1963), *Biochem. J.*, **88**, 452–459.

Richmond, M. H. (1975a), *Methods in Enzymology*, **43**, 664–672.

Richmond, M. H. (1975b), *Methods in Enzymology*, **43**, 672–677.

Richmond, M. H. and Sykes, R. B. (1973), *Adv. Microbial Physiol.*, **9**, 31–88.

Rolinson, G. N. (1971), *Proc. R. Soc. London Ser. B*, **179**, 403–410.

Rolinson, G. N., Batchelor, F. R., Butterworth, D., Cameron-Wood, J., Cole, M., Eustace, G. C., Hart, Marian V., Richards, M. and Chain, E. B. (1960), *Nature*, **187**, 236–237.

Ross, G. W. (1975), *Methods in Enzymology*, **43**, 678–687.

Ross, G. W. and O'Callaghan, C. H. (1975), *Methods in Enzymology*, **43**, 69–85.

Sakaguchi, K. and Murao, S. (1950), *J. Agric. Chem. Soc. Japan*, **23**, 411.

Savidge, T. A. and Cole, M. (1975), *Methods in Enzymology*, **43**, 705–721.

Sawai, T. and Lampen, J. O. (1974), *J. Biol. Chem.*, **249**, 6288–6294.

Sutherland, R. and Rolinson, G. N. (1964), *J. Bacteriol.*, **87**, 887–899.
Sykes, R. B. and Richmond, M. H. (1970), *Nature*, **226**, 952–954.
Thatcher, D. R. (1975), *Methods in Enzymology*, **43**, 640–664.
Vandamme, E. J. and Voets, J. P. (1974), *Advances Appl. Microbiol.*, **17**, 311–369.
Vandamme, E. J., Voets, J. P. and Dhase, A. (1971), *Ann. Inst. Pasteur*, Paris, **121**, 435–446.
Vanderhaeghe, H., Claesen, M., Vlietinck, A. and Parmentier, G. (1968), *Appl. Microbiol.*, **16**, 1557–1563.
Waldschmidt-Leitz, E. and Bretzel, G. (1964), *Z. Physiol. Chem.*, **337**, 222–228.
Walton, R. B. (1964), *Develop. Ind. Microbiol.*, **5**, 349–353.
Wolfe, S., Godfrey, J. C., Holdrege, C. T. and Perron, Y. C. (1963), *J. Amer. Chem. Soc.*, **85**, 643–644.
Yamamoto, S. and Lampen, J. O. (1975), *J. Biol. Chem.*, **250**, 3212–3213.

Chapter 6

Patenting Developments with Micro-organisms and their Products

F. S. M. GRYLLS, Distillers Company Ltd, Morden, Surrey

6.1 INTRODUCTION

In order to understand the philosophy of the patent system, it is worth while glancing back to the historical origins of today's Patent Law in the United Kingdom.

Patents granted for inventions are only one example of Letters Patent, that is to say of Letters issued by the Crown and addressed to the Crown's subjects. Such Letters Patent are the normal form of making a variety of special grants of privilege, for example to offices of State and the like, including monopoly rights.

In medieval times, monopoly rights were granted by the Crown both to trade guilds and individuals, and were much used to protect, for example, various crafts and trades.

One use to which such monopoly rights were put was to encourage the creation in the United Kingdom of new industries imported from Europe. Such industries could be set up by granting the foreign craftsmen exclusive rights in their trade for a set period of time, sometimes in exchange for training apprentices in their particular skill.

The system was subject to a great deal of abuse, for many monopolies were granted for the simple benefit of the recipient in such matters as trade, etc. Over the years, attempts were made to stamp out these abuses, and these culminated in the year 1610 in the proclamation referred to as *The Book of Bounty* in which was stated that, *inter alia*, monopolies were one of the 'special things for which We . . . command no suitor presume to move Us'. Nonetheless, projects for new inventions, recognized as beneficial to the public, were excepted from this prohibition. The philosophy behind this exception is to be found in a Court pronouncement made in 1615 as follows:

'But if a man hath brought in a new invention and a new trade within the kingdom in peril of his life and consumption of his estate or stock, etc., or if a man hath made a new discovery of anything, in such cases the King of his grace and favour in recompense of his costs and travail may grant by charter unto him that he shall only use such a trade or trafique for a certain time, because at first people of the kingdom are ignorant, and have not the knowledge and skill to use it. But when the patent is expired the King cannot make a new grant thereof.'

It is interesting to observe that even to this day under current United Kingdom patent law, an invention is regarded as novel provided it has not been published or used within the United Kingdom before the Patent Application is lodged.

In 1628 the Statute of Monopolies was passed, again attempting to tighten up against the abuse of monopolies, and reciting in Section 6 what is commonly accepted as the foundations of present patent law. This Section runs as follows:

'Provided also (and be it declared and enacted) that any declaration before mentioned shall not extend to any Letters Patent and grants of privilege for the term of fourteen [now sixteen] years or under, hereafter to be made, of the sole working or making of any manner of new manufactures within this realm, to the true and first inventor and inventors of such manufactures which others at the time of making such Letters Patent and grants shall not use, so as also they be not contrary to the law or mischievous to the State, by raising prices of commodities at home, or hurt of trade, or generally inconvenient.'

It will be seen, therefore, that the concept of granting limited monopolies for inventions is that the inventor should be rewarded for his endeavour, by virtue of a time limited monopoly, but that the public too should benefit for the grant of this privilege by disclosure of the invention, followed by the eventual right of the public at large to work the invention after the monopoly rights have expired.

The Patents Acts in force today stem directly from these early days, and carry the same underlying philosophy, namely that to encourage the development of industry in the Kingdom, the inventors may be granted temporary monopoly rights, provided that they disclose the manner of working their invention to the public. The alternative to such temporary grant of monopoly rights would be for the inventor to keep his methods secret, and, insofar as this in practice can be very difficult, the result would probably be to discourage a substantial amount of research and development activities. That at least is the theory.

Patents will only be granted for inventions as defined by the Acts, the definition being 'any manner of new manufacture . . . and any new method or process of testing applicable to the improvement or control of manufacture . . .'. Generally speaking, the invention must relate to some tangible process or product and cannot relate to theories. Generally speaking, patents will not be granted for medical or agricultural processes which, until recently, have not been regarded as industrial matters. However, there is to be seen a slight shift in the interpretation of the law in this respect, the cause of which will be referred to later.

In order to obtain grant of a patent, it is necessary for the inventor, or his legal assignee, to apply to the Patent Office, usually through a Patent Agent, in the appropriate manner. This involves lodging in the first instance what is known as a Provisional Application, in which is set out the outline of the invention and its method of operation, as appropriate. A priority date will be assigned to this document, which will be the date of receipt at the Patent Office. This priority date is most important, and for this reason Applications for patents should be filed as swiftly as possible after the invention has been made.

Following lodging of the Provisional Patent Application, the applicant must follow up, within twelve months of the priority date, with the Complete Specification. This document will set out in detail the nature of the invention, will include working examples (typically drawn from a laboratory notebook) setting out all the procedures used in great detail, and will end with the Claims. The Claims are of crucial importance, as they will define the scope of the monopoly requested, and it is around the Claims that future battles to prevent people working the invention, or to grant licences for working the invention, will revolve. Drafting this document, and indeed the Provisional document, is of such importance that it is unwise for the amateur to attempt it for himself.

On receipt of the Complete Specification—which incidentally can be filed at any time up to twelve months after the Provisional document, and, indeed can be filed initially in place of the Provisional document—an Examiner in the Patent Office will check the Application for a number of formal requirements relating to format and minor legalities. His most important duty, however, is to check the Application for novelty. That is to say the Examiner must ensure, insofar as is possible, that the invention claimed by the inventor has not been published elsewhere. The Examiner can use any literature published in the United Kingdom, but he must at the very minimum examine previous United Kingdom patent Specifications in the same field. He will further check that there is no pending or unpublished Specification within the Patent Office relating to identical subject matter. Provided that the applicant overcomes all these hurdles, then the Examiner will accept the Patent Application and in due course the Complete Specification will be published.

These published Specifications are advertised in the weekly Patent Office Journal, and may be inspected or purchased from the Patent Office by any interested member of the public. An interested party can give formal notice to the Patent Office that he objects to grant of a patent based upon the published Patent Specification which has previously been accepted by the Examiner. This system enables the public to draw to the attention of the Patent Office matters relating to the invention which were not known to the Examiner. For example, there may be some highly relevant literature disclosing the invention which was not found by the Examiner, or there may have been some public use of the invention of which the Examiner was totally unaware. Alternatively, it may be that a member of the public will allege that the invention properly belongs to him, and was obtained, or stolen, from him by the applicant. Altogether, there are eight different headings under which an objector could raise opposition proceedings, but the most common relate to lack of novelty or anticipation, and what is known as 'obviousness'. Obviousness, as a ground of rejection of a patent Specification, is different from novelty in the sense that the invention, although admittedly novel, is nonetheless a clear and obvious development of known prior art. For example, following the general availability of stainless steel, it would be perfectly obvious to make table cutlery from this metal. Nonetheless, such cutlery would be strictly novel in the sense that it had not been made before. A Patent Application claiming stainless steel table cutlery could and would be successfully opposed by a third party on the ground of obviousness.

Another major ground of opposition relates to what is known as 'sufficiency'. As explained before, the *quid pro quo* which the Crown demands in return for granting a monopoly for a limited period of years to an inventor, is that the inventor should disclose his invention to the public. It is not uncommon for a Patent Application to be published which is not sufficient in descriptive content to enable a person skilled in the art to work the invention. In such a case, the third party would be entitled to object to grant of a patent on the published Specification on the basis that the inventor has not revealed his invention. For example, it may be that the inventor has claimed a new type of catalytic reaction for the oxidation of hydrocarbons. He may have disclosed his catalyst and the general nature of his reaction, but may have deliberately omitted to disclose that the reaction would only proceed satisfactorily under a very narrow band of temperature and pressure conditions. By not revealing these special conditions required, the inventor would be attempting to obtain a monopoly right without, in return, handing over his invention to the public. It would be quite legitimate, therefore, in such cases to object to the grant of a patent on the grounds of insufficiency of description.

It should perhaps be pointed out here that, in considering prior publications as anticipating a Patent Application, *any* document published in the U.K. can be used. What is not commonly appreciated is that this embraces publications made by the inventor himself. In other words, if the inventor has published a description of his invention before he has filed an Application for a patent, then his own publication can be used as a cited prior document. It is therefore quite imperative for any inventor to maintain total secrecy concerning his invention until such time as he has obtained a priority date for a Provisional Patent Application filed in the Patent Office.

Reverting back to the publication of the accepted Patent Specification, if no member of the public comes forward objecting to grant, then, after three months following publication of the Specification, the patent will proceed to grant. A fee will be payable by the applicant, and the granted patent, when successfully through this last formal stage, is now truly Letters Patent.

Nonetheless, these granted Letters Patent, which will remain in force subject to payment of annual renewal fees for sixteen years following the date of filing the Complete Specification, are not themselves indestructible. Just as prior to grant it was possible for a third party to object to grant on a number of grounds so, following grant, can an action be brought in the High Court for revocation of the granted patent on any one of a number of grounds similar to those used in opposition proceedings. The difference between an opposition and a revocation action is that the former is carried out within the Patent Office, and is a relatively cheap and simple procedure, but the latter, because the patent has now been granted, is carried out before the High Court, and is relatively costly. There are subtle differences in practice between the grounds available for opposition and the grounds available for revocation, and frequently it is considered better not to oppose at the pre-grant stage but to await grant and take up a revocation action.

What is important to remember is that, in spite of grant of a patent, it does not necessarily follow that the patent is valid. Thus, if the owner of a patent uses his rights to prevent a third party from infringing, i.e. from working his invention, then it is always open for the third party to challenge the validity of the patent. In this respect, the Crown grant of patent rights is unusual insofar as this Crown grant can be challenged.

Now let us take stock of the position. Our inventor, following Application at the Patent Office, following examination of his Patent Application by the Examiner, following acceptance and publication of his Patent Application, leading to grant of his patent (assuming no third party successfully challenged in opposition proceedings), now has a document called a Letters Patent, to which will be assigned a serial number. He is now theoretically at least in a position to prevent third parties operating his invention. He may for example have invented a novel type of ball point pen, and he may wish to manufacture this and exploit his invention in this manner. Other manufacturers of pens would not be permitted to copy his invention and, if they did so, then the inventor would be entitled to sue for infringement. This can be an expensive business and is not lightly to be undertaken. The exact nature of the infringing article will have to be compared very carefully with the Claims for the invention as set out in the Patent Specification, and at all stages expert advice will have to be sought. Also, as mentioned earlier, the alleged infringer, as one defence, would undoubtedly sue for revocation of the Patent Specification if he had sufficient grounds. Generally speaking, excepting in major issues, the mere existence of the Patent Specification tends to keep competitors 'off the grass'. It is not common for infringement actions relating to minor matters to come before the Courts: such cases are usually settled by consent. The major cases generally involve honest differences of opinion as to whether the alleged acts really infringe the Claims of the Patent Specification if he had sufficient grounds. Generally speaking, be (1) that the acts of the defendant do not infringe the Claims; (2) if they do infringe the Claims, generally the Claims themselves are invalid.

Quite often, however, the inventor does not necessarily want to work the invention for himself. He would rather see his invention developed commercially by a third party, and, in such cases, the inventor will arrange to license the third party to operate under the Claims of his Patent Specification. Probably the greatest value of most patents lies in the ability to license the rights under them to other manufacturers. License royalties form a significant part of the revenue of many major companies, particularly in the chemicals, plastics, and pharmaceuticals field. The owner of the patent can give an exclusive licence to a single manufacturer, or he may if he wishes licence as many people as is commercially desirable. He may put geographical constraints upon the licence, and also certain sales constraints. He may grant a licence to a third party, whilst still retaining the rights himself to manufacture the articles according to the invention, or he may undertake not to enter manufacturing operations at all. All of these arrangements are within the domain and rights of the owner of the patent.

However what he may not do is abuse his monopoly rights. Contrary to popular public opinion, the owner of a patent cannot, for example, refuse both to work the invention himself and refuse to grant licences to third parties. Provided it can be shown that there is a public need for the invention to be used or made available, then the patentee, if he refuses himself to use the invention, can be compelled eventually to grant licences to third parties. The various stories you have all heard, of the large petrol company which bought up the patent rights in carburettors giving extended mileage, in order to suppress the invention, are quite false; as indeed are all those stories about the match companies who have purchased patent rights in everlasting matches. Should such a situation occur, that is to say should monopoly rights thus be abused, then a third party manufacturer can, under various provisions of the Patents Acts, apply to the High Courts for the grant of a compulsory licence. Indeed, in the case of inventions relating to food and medicine, a third party manufacturer is entitled, as of right, to obtain a licence even in competition with the patentee himself.

The third party manufacturer would have to pay royalties to the inventor or the owner of the patent, but nonetheless it is not possible for such inventions to be held exclusively by the owner of the patent.

Likewise, in granting a licence, the patentee is not able to make excessively burdensome conditions. For example, if the invention required as starting material sulphuric acid, then the owner of the patent would not be allowed to make it a condition of giving a licence that the licensee had to purchase sulphuric acid from the patentee. He could only do this legitimately if, at the same time, he offered an alternative licence in which such a provision was not included. This would then give the licensee the freedom to buy his raw materials wherever he wishes.

It will be seen therefore that the Crown, whilst granting a monopoly right to an inventor as fair reward for his endeavours, also protects the interest of the public at large by various provisions in the Patent Act which are designed to prevent the inventor from abusing his monopoly, either by preventing the public access to the invention under reasonable terms and conditions, or by refusing to work his invention himself. Incidentally, a further example of the limited nature of the monopoly granted is that experimental use of a patented invention is not an infringement. In other words, the Crown recognizes the importance of the freedom to experiment in order to further industrial progress. However, once commercial use is involved, then infringement will occur, and a licence must be arranged. Much of the criticism of the patent system is thus both ill-informed and incorrect.

This general and introductory account of patent law is, of course, directed to the United Kingdom law. However, much of the principles outlined above apply to other countries, with minor variations, although in many countries there are no provisions for automatic rights to a licence for inventions relating to food and medicines. The more normal provision abroad is that, after three years' non-use of an invention, then a compulsory licence can be sought.

Turning to the question of the priority date, this you will recall is the date upon which the Provisional Patent Specification was lodged in the Patent Office. It is necessary to file your Complete Specification within twelve months from that date. Now it may be that the inventor or the applicant (not necessarily the same person) may wish to seek patent protection abroad as well as in the United Kingdom. In this event, he can lodge an Application and can claim, under an international agreement, the priority date of his original Provisional Patent Application. In order to do so he must file his foreign Application within twelve months of filing the Provisional Application in the United Kingdom. Likewise, a foreign applicant in the United Kingdom can use the priority date of his first Application made in his own country, provided he files in the United Kingdom within twelve months of that date. Thus, by international agreement, the priority date of the first originating Patent Application is available in any other country (by and large) provided (1) evidence is produced showing how the original priority date was arrived at, and (2) the foreign Application is made within twelve months of that date.

Taking out foreign Applications can be a very expensive and frustrating experience. Particularly is this true in those cases where the future commercial viability of the invention is difficult to forecast. Each country has its own particular rules and regulations, and one of the greatest costs involved is in arranging for translations into the native language of the country concerned; this can be a particularly expensive operation. However in the last few years a great deal of work and effort has been put into a series of international deliberations which, in the not too distant future, should culminate in facilities being set up, probably in Munich, for a single Patent Application to be filed, leading in turn to grant of patents in a nominated list of European countries. Only one examination will be involved. A further development will be for a single patent to be granted which will cover all the members of the E.E.C. However, these are things for the future, and for some considerable time national patent systems of each country will run in parallel with these new type of proposed patents.

6.2 MICRO-ORGANISMS AND THEIR PRODUCTS

With this general background of patent law in mind, it is opportune now to turn to specific matters relating to the microbiological industries, and more particularly to the current problem relating to the nature and timing of deposition and release of cultures, when such cultures form an essential part of the claimed invention, with some consideration also of the overall question as to whether or not microbiological processes are proper subject matter for patents. In considering these problems, it is necessary always to bear in mind that the fundamental purpose of the patent system is to provide the inventor with a time-limited monopoly in exchange for his undertaking to disclose his secrets in such a manner that, on expiry of his monopoly, the public may freely work his invention.

The deposition problem—or alleged problem—because it does not really exist at all, is stated simply thus: if a culture is vital to the operation of an invention:

(1) Should there be a clear requirement that the culture must be deposited in a recognized collection and, if so, when should the deposition take effect?
(2) Upon what terms and conditions, including timing, should the culture so deposited be made available to the public?

There is a developing feeling, at least on the official side of the fence, that the requirements should in the foreseeable future be:

(a) The culture should be deposited on or before the priority date (or the date of completion).
(b) The culture should be publicly available as soon as the patent application is open to public inspection. This availability should be subject to certain reservations and restrictions.

In the United Kingdom, following the House of Lords Appeal in Dann's Application, the present law appears to have become crystallized, to the effect that deposition or release is not necessary. In American Cyanamid versus Berk, Mr Justice Whitford at one stage, of legal necessity, went along this finding, but later on found a neat way around this precedent, as will be described later.

It is contended that much of the current discussion, together with the decision referred to, is irrelevant; and further that, at least in the United Kingdom, the present law (provided one could eliminate the case law precedents of the House of Lords) clearly and properly sets out the correct requirements. To support this sweeping generalization, it is helpful to look backwards through the centuries into history to remind ourselves what is the underlying philosophy of patent protection.

Patents, as stated earlier, are in effect time-limited monopolies granted by the Crown. Insofar as the Crown historically does not generally favour monopolies, it is clear that monopolies for inventions have been put in a separate category as being considered to be more beneficial to society than harmful, but there has always been a clear intent to limit the time of the monopoly, and a clear approach that a limited patent monopoly is a fair reward to the inventor for his efforts, but thereafter his invention shall be available for general use.

That was the intent, and still must be the intent. The price for the monopoly must be paid, and must be paid in advance. The payment comprises revealing to the public the new invention in such manner that, on expiry of the monopoly (or even before), the public can freely and readily work the invention for themselves (by 'public' is meant, of course, persons skilled in the art). Therefore, in order properly to be granted his monopoly, the inventor must make a detailed and full disclosure of his invention to the public, in whatever manner necessary to enable the invention to be worked.

This then should be the backcloth to the various arguments and points of view put forward concerning conditions under which microbial processes should be patented.

In similar manner, arguments relating to whether or not such microbial processes should even be eligible for patent protection are clarified by looking back into history. As every person interested in patents knows, 'invention' was originally defined as 'any manner of new manufacture', and the definition of 'invention' today under Section 101 of the Patents Act, 1949, contains the same phrase, together with the extension that 'invention' includes any new method or process of testing applicable to the improvement or control of manufacture.

Over the centuries many arguments have been raised as to what is or is not a manner of new manufacture. In the opinion of the author, these arguments in the main are either spurious or, at best, unnecessary. In the days when 'invention' was defined in the words quoted above, there was little thought or possibility that inventions or developments could relate to other than what one might call 'inorganic processes'. The prevailing concept would surely have been that all matters involving life, be they agricultural or biological, were Divine matters upon which mankind could have no influence. It is therefore probable that in the original definition of 'invention' there was no conscious attempt positively to exclude inventions in the biological spheres from being patented; it was simply that these sort of possibilities did not cross man's mind in those days. The sole purpose of the words so chosen was, almost certainly, to indicate that the Crown, whilst frowning upon restrictive practices and monopolies generally, would nonetheless grant monopolies for new and useful activities which, by virture of their progressive and ingenious nature, would benefit society generally.

Nonetheless, as the history of patent law unfolded throughout generations of case law, and as mankind's inventive ingenuity became such that it spilled over into the biological arts, we find an approach to patentable subject matter developing which appears to be misguided, and attempted to read far too much into the purpose behind the original definition of 'invention'. As stated above, there appears to be no valid reason for the belief that the original definition was designed to exclude biological inventions; but, nonetheless, as such inventions became possible, successive courts re-confirmed their belief that the original phrase was carefully chosen, almost with supernatural insight into twentieth-century biological developments, and thus we arrived at the position earlier this century where essentially all biological subject matter was precluded from patent protection. However, pressures built up on the courts as successive applicants challenged this misguided concept. In consequence the courts, by the force of circumstances, were almost compelled, insofar as they were entitled to, to reconsider and attempt to distinguish from previous adverse judgments.

Thus we find cracks starting in the firm front presented against allowing patents for agricultural processes, and treatment of humans or animals. For example, propagation of mushroom tissue in deep culture (Szuecs' Application) was considered to be patentable on the ground that the propagation methods specified in the patent differed entirely from the conditions appertaining under natural conditions. On the other hand, a sophisticated and highly artificial greenhouse technique for forced growth of Poinsettia (Philips' Application) was refused on the distinguishing grounds that the plant, albeit under unnatural conditions, was merely manifesting its natural, and presumably Divinely-given properties. Clear-cut agriculture processes were still too 'near the bone' for the courts to accept. The truth of the matter was that the pattern had been set some time earlier when Dr Weizman invented his process for the microbiological production of acetone, and the court, with unaccustomed lack of prevarication, clearly stated that 'in such an invention there is patentable subject matter' (43 R.P.C., 185). Thus the invention relating to use of bacteria was allowed because, in the context of its time, it could safely be classified as a manner of new manufacture without, as it were, treading on the toes of the previous attitudes concerning agricultural processes and the like. This in turn must have had considerable influence in permitting Szuecs' Application for mushroom culture to be allowed, which then led inevitably to an arbitrary distinction being drawn, with little justification, between this type of biological case and the Poinsettia case.

It appears therefore that in order to enable this whole new field of microbiological technology to be regarded as patentable, without discarding the long held but wrongly held concepts regarding the non-patentable nature of biological processes, we have been led to the delusion that indeed microbiological processes represent a distinct and unique class of invention, and, therefore, in some way require special treatment.

On the contrary, it is totally wrong to regard microbiological processes as something special, and the patent law should not attempt, either by case law or by statute law, so to do. To arrive at this conclusion it is necessary to go back to first principles and reconsider, as earlier described, what patents are all about. They are all about giving a time-restricted monopoly to a diligent inventor, and the *quid pro quo* of the inventor is to reveal his secrets to the public at large so that, after the time limit has expired, they may benefit freely from his invention, and incidentally also in the meantime use the invention freely for experimental purposes. If that objective is clearly held in sight, then much of the heat and dust that has been recently generated concerning the release or otherwise of cultures in microbiological inventions, and their overall patentability, would vanish immediately.

As mentioned earlier, the questions asked are these:

(a) Should the inventor at any time be compelled to release his culture to the public? and
(b) If the inventor should so be compelled, what should be the exact timing and mode of release?

The argument generally presented by inventors runs thus: if the invention hinges around a new culture, e.g. for the production either of a new antibiotic or the more efficient production of a known antibiotic, then it is unfair to insist that the applicant for a patent must deposit his culture in such a manner that it is freely available to the public, and quite unfair that he should have to do so before he has been ensured grant of a patent on his application.

This objection is surely invalid. The total disclosure of the invention in return for the limited monopoly is the contract. This is a bargain which the Applicant knowingly enters into, and to plead that because his invention is a magic culture, that makes him different from all other categories of invention, is really quite absurd. Opponents of this view will counter-argue that, if the culture has to be made available prior to the grant, and grant is subsequently refused, then the inventor has handed over his invention to the public with no recompense. This indeed is the situation, but how is this any different from any other form of invention? In every other case, certainly under United

Kingdom jurisdiction, the inventor is compelled to disclose his invention on making application, and eventually, several months before grant, his written description of his invention is laid open to public examination. If grant is subsequently refused, then the inventor has lost his secrecy and secured no monopoly and this, at least in theory, is the situation regarding all patent applications which are accepted, but upon which grant is subsequently refused.

Why, therefore, should the holder of a culture be specially privileged in being enabled to conceal his invention from the public? It may be argued, and indeed is, that many patent applications today successfully conceal the method of operating the invention. That is to say, the Specification is drafted in such general terms that even with the Specification to hand the public cannot optimize or sometimes even work the invention. The remedy then is to tighten up the procedures which have allowed this situation to develop, for failure to disclose the invention in the Specification is quite contrary to the law, vide Section 4(3) of the Patents Act, 1949:

'Every complete specification:

(a) shall particularly describe the invention and the method by which it is to be performed;
(b) shall disclose the best method of performing the invention which is known to the applicant and for which he is entitled to claim protection; and
(c) shall end with a claim or claims defining the scope of the invention claimed.'

The present discontent concerning inventions relating to cultures should not be regarded as reasons for making special provisions to enable the inventor to maintain secrecy whilst enjoying monopoly rights, but rather should focus attention upon all those other patents wherein the Applicant has wrongly but successfully managed to acquire his monopoly rights without proper disclosure. It is difficult to see how this argument can be denied without simultaneously challenging the whole concept of the patent system. That is not to say that it is wrong to challenge the total concept of the patent system, but merely that it is wrong to make what are no more than special pleas for secrecy regarding a narrow field of inventions.

It will be seen how the law has stood upon its head, and how present arguments seem to be directed at distorting the whole concept of the patent system if we examine some recent statements and pronouncements.

Firstly, let us study the Consultative Document on Patent Law Reform now circulating, and published by the Department of Trade. This document supplements the White Paper on Patent Law Reform, and sets out in detail the Government's proposal for implementing the recommendations of the Banks Committee, and for enabling the United Kingdom to ratify the various European Patent Conventions and the Patent Co-operation Treaty now under negotiation. Paragraph 58 (page 17) states:

'For the specification to provide a sufficient disclosure of the invention, a culture of the micro-organism must have been deposited in a recognised culture collection not later than the filing date of the application, the application as filed must contain such relevant information as is available to the applicant about the characteristics of the micro-organism, and the application must, within two months of filing, identify the culture collection and give the date and the file number of the deposit.' So far so good. 'As from the date of publication of the application, the culture deposited is to be available to any person on request addressed to the culture collection and containing identification of the person making the request, an undertaking *vis-à-vis* the applicant or patentee not to make the culture available to anyone else and, where the request is made before grant, an undertaking *vis-à-vis* the applicant to use the culture for experimental purposes only.'

These special restrictions have been proposed no doubt because of the currently prevailing arguments that microbiological inventions are special cases. Consider the case of an applicant for an invention which is not in this so-called special field, e.g. the case of an inventor who has invented an improved safety lock. The applicant will quite properly have fully disclosed and described his invention in his Complete Specification, and will have appended thereto proper drawings. In the fullness of time, this Specification will be published. Is he to be enabled to demand, pending grant, that nobody can buy a copy of the Specification unless he gives his name and address in advance, promises not to let his wife and children read the document, and further promises to consume or otherwise destroy the Specification the moment he has read it? If not, why not? Why should the applicant for an invention concerning micro-organisms be put in a special privileged position? There is no convincing reason whatsoever for this special provision to meet so-called special cases.

It is quite astonishing that anybody can really believe that the disadvantages of publication of inventions are special and peculiar to the owners of micro-organisms. They apply with equal force to many other fields of industry, and, if it is considered that the law is harsh in this respect, then pleas should be made to have the law amended. Pleas should not be made for a particular area of inventive activity to be the recipient of favouritism. Nonetheless, these statements seem to represent the prevailing and developing attitudes which are viewed by some with concern.

Two important decisions which have generated so much discussion in the last year or two must now be referred to, merely to illustrate the point that the essence of the matter is, or should be, whether or not the applicant has paid the proper price of full disclosure in return for his temporary monopoly.

When Dann's Application reached the House of Lords, their Lordships, or certaily four of them, spent a tremendous amount of time in abstract argument concerning the finer points of patent law. Their deliberations, however, appear to be wholly irrelevant to the real issue of the day, which was whether or not Dann's Application could be a valid grant in spite of the fact that the cultures used in the microbiological procedures were essential to the operation of the process, but were not available either at the time of filing, or publication or grant. Some of their Lordships became enveloped in fascinating dissertation as to whether the Patent Acts unequivocably demanded the deposit of cultures or not. Naturally they were obliged to conclude that the Acts did not specifically call for this requirement, primarily on the grounds that the present Acts do not call for general deposition of any specimens, samples, intermediates, and the like as a mandatory requirement. Apparently their Lordships never addressed themselves to the real question, which surely should have been 'had the applicant on publication paid the full price, namely the price of letting the public work his invention, experimentally during the life of the patent, and commercially thereafter?' The answer to this question is clearly 'no', and the argument should surely have stopped there without more ado. It is symptomatic of the artificial approach of these cases that the same four Lords studiously avoided getting involved in the question as to whether or not microbiological processes in any event are fit subject matter for patents. Lord Diplock did however address himself to the real points at issue.

He argued convincingly and logically that, once one has extended the concept of 'invention' to include new antibiotics resulting from the discovery in nature of the parent micro-organism, the meaning to be ascribed to the provisions of the Patent Acts must take account of that extended concept. It followed, therefore, he argued, that the requirement in Section 32(1) (h) of the Acts, that the Complete Specification must 'sufficiently and fairly describe . . . the method by which it (i.e. the invention) is to be performed' as one which is not satisfied unless the Specification states where a culture of the parent micro-organism can be obtained by the reader. Without that information, he concluded, the invention cannot be performed by the reader at all. Without access to the culture, there is no counterpart for the patentee's monopoly. The provision of this counterpart constitutes a part of the basic policy of the Act.

In plain language, what is surely being stated here is that, once it has been agreed that microbiological inventions are proper subject matter for patents under the existing Patent Acts, they must quite naturally fulfil the requirements of those Acts as they stand, and should not be treated as a special or privileged category of inventions. In order to fulfil the requirements the inventor must enable the public to work his invention and, if this means parting with his culture, then so be it.

Turning now briefly to American Cyanamid versus Berk, but bearing in mind that this case has not been to the House of Lords, it is gratifying to see in print some statements attributed to various parties which support the view that microbiological inventions are not special cases requiring special privileges.

In the introductory comments of his judgment, Mr Justice Whitford drew attention to the necessity not to be hide bound in attempting narrow definitions of 'manner of new manufacture'. This was in relationship to one of the objections to the patent by Counsel, namely that biochemical processes were not fit subject matter for patent protection (this case related to the alleged infringement of American Cyanamid Company's patent, which claimed a novel process for the production of tetracycline using certain strains of Streptomyces). Mr Justice Whitford indeed quoted Lord Diplock in Dann's Application, and I can do no better than quote the quote as it were (F.S.R., 1973, page 509):

'The only definition of "invention" contained in the Patents Acts, 1949, is by reference to the phrase "any manner of new manufacture" in the Statute of Monopolies 1623. This statute has never been construed as confining the grant of patents to processes which would have been within the contemplation of Parliament in the early seventeenth century as constituting a new manner of manufacture. The concept of "invention" as an activity which entitles the person who undertakes it to a patent monopoly of a process or its product has been continually changing by applying the social policy which underlay the Statute of Monopolies in the seventeenth century to the changing conditions resulting from advances in technology and the sciences. That policy was to provide a material inducement to persons to undertake whatever effort and expense might be needed to introduce into the United Kingdom some useful new process or new product with a view to its becoming generally available for use or manufacture by the inhabitants of the realm. The material inducement provided was the right for a limited period to prevent other persons from using the new process or product in the United Kingdom without the consent of the inventor; but the object of the policy would not be achieved unless at the end of that period those persons wishing to engage in the business of operating processes or making products of kind were provided with all the information necessary to enable them to operate the process or to make the product.

'The counterpart of the temporary monopoly granted by the patent was the provision of the information. I use the word "counterpart" rather than "consideration". This word has acquired a technical meaning in English law whereas the basic policy which underlies the Patents Act, 1949, has been adopted by many other nations which provide corresponding monopoly to the inventor within their territories under their own patent laws.

'It is in my view consistent with this basic policy to treat the kind of research involved in the discovery of a strain of micro-organism from which a new and useful antibiotic can be prepared as an activity which entitles the person who undertakes it to a temporary monopoly under the Patents Act, 1949, of the product of his success. I accept, therefore, the extension of the concept of "invention" to include antibiotics which are discovered through this kind of research.'

Mr Justice Whitford further quoted from N.R.D.C.'s application in Australia relating to a method for eradicating weeds. The Chief Justice of that country stated the following (F.S.R., 1973, page 511):

'The inquiry which the definition demands is an inquiry into the scope of the permissible subject matter of letters patent and grants of privilege protected by the section. It is an inquiry not into the meaning of a word so much as into the breadth of the concept which the law has developed by its consideration of the text and purpose of the Statute of Monopolies. One may remark that although the Statute spoke of the inventor it nowhere spoke of the invention; all that is nowadays understood by the latter word as used in patent law is comprehended in "new manufactures". The word "manufacture" finds a place in the present Act, not as a word intended to reduce a question of patentability to a question of verbal interpretation, but simply as the general title found in the Statute of Monopolies for the whole category under which all grants of patents which may be made in accordance with the developed principles of patent law are to be subsumed. It is therefore a mistake, and a mistake likely to lead to an incorrect conclusion, to treat the question whether a given process or product is within the definition as if that question could be re-stated in the form: "Is this a manner (or kind) of manufacture?" It is a mistake which tends to limit one's thinking by reference to the idea of making tangible goods by hand or by machine, because "manufacture" as a word of everyday speech gradually conveys that idea. The right question is: "Is this a proper subject of letters patent according to the principles which have been developed for the application of section 6 of the Statute of Monopolies?".'

This indeed is a remarkably clear exposition of the logical approach to this problem.

Let us turn now to what Mr Justice Whitford concluded concerning the availability of cultures in such cases.

In the first instance, of course, he was severely handicapped in arriving at a logical judgment because of the precedent set by their Lordships in Dann's Application. Nonetheless, it appears that Mr Justice Whitford managed at the same time whole-heartedly to agree with their Lordships on the question of *availability*, while simultaneously essentially stating that the patent in question was invalid because nobody knew *what* the cultures were. Their Lordships in Dann's Application had said it didn't matter at all if nobody knew *where* the cultures were. Mr Justice Whitford said 'I fully agree, but it does matter if you don't know *what* they are'. In compulsorily concurring with their Lordships on *where* the cultures were, Mr Whitford said on the question of availability (F.S.R., 1973, page 521):

'The question of availability must be left out of consideration completely. What, as I understand it, the House of Lords held was that, irrespective of availability, the duty still rests on the patentee of providing an adequate and sufficient description in the body of the specification. The submission of counsel for the defendants, which I think is correct, was that it was incumbent upon the plaintiffs, the applicants, as at the date of the application, to describe the novel mutant strains of the invention sufficiently well to identify them to other workers in the field. To do this, he said, it was their duty to do the best they could to provide as accurate a description as was possible, using, for the purposes of such description, the criteria which had become accepted by taxonomists in 1962 to define micro-organisms within this particular species.'

The Judge thereafter, concluded, that the description of the micro-organisms in the patent Specification was insufficient to enable them to be identified, and he added that, *inter alia*, the patent must fail on this ground. Thus we arrive at a fair and reasonable solution, which should have been obvious all along.

This tortuous path back to basic principles was only necessary because words rather than principles had distracted the attention, and continue to distract the attention of those involved, for far too long.

To conclude therefore, let me re-state the timeless principle yet again—an inventor may, if he so wishes, seek the Crown's assistance in obtaining a time-limited monopoly for his new and useful invention if, and only if, he hands over his secrets for eventual public use.

Chapter 7

Industrial Glucose Isomerase

Dr C. BUCKE, Tate & Lyle Ltd, Group Research and Development, University of Reading, Reading, Berkshire

7.1 INTRODUCTION

Wiseman (1975) wrote 'Glucose isomerase may be the most important of all industrial enzymes of the future.' It is already 'big business' and it will get even bigger: the estimated production of high-fructose glucose syrup (HFGS) in the United States in 1975 was 1,400–1,800 million pounds and in 1977 production should be 3,500 million pounds (*Reuter Sugar Report, 16th April 1975*). The estimated potential market for HFGS is 4,500–6,000 million pounds per annum and the present usage is limited only by its availability. Plant for the production of HFGS is under construction in the United Kingdom and in the German Federal Republic and small amounts of this product are already being produced in Belgium. Glucose isomerase (GI) is therefore of world-wide significance. This chapter will describe the discovery and development of the enzyme and the technology of its production and use. It is inevitable in such a rapidly-expanding area of research and commerce that some important advances have been shrouded in secrecy; and therefore there are tantalizing gaps in the academic and patent literature relating to some of the more significant processes. The academic and patent literature has been studied up to mid-November 1975.

7.2 THE ENZYMES AND THEIR PROPERTIES

7.2.1 Introduction

The conversion of glucose to fructose is one of a group of reactions collectively known as the Lobry de Bruyn-Alberda van Ekenstein transformation (Figure 7.1) (see Speck, 1958). Such reactions are favoured by alkaline conditions and high temperatures and the alkaline isomerization of glucose has been studied in some detail as a possible means of producing fructose commercially (e.g. Scallet and Ehrenthal, 1967, 1968; Scallet *et al*., 1972; Barker *et al*., 1975). However, the lack of selectivity of the alkaline isomerization allows the production of non-metabolizable materials such as psicose (Figure 7.1) and objectionable, coloured materials which are costly to remove (MacAllister *et al*., 1972). Consequently, chemical isomerization of glucose has not been employed commercially.

It is obvious that a specific method of performing this reaction, for instance

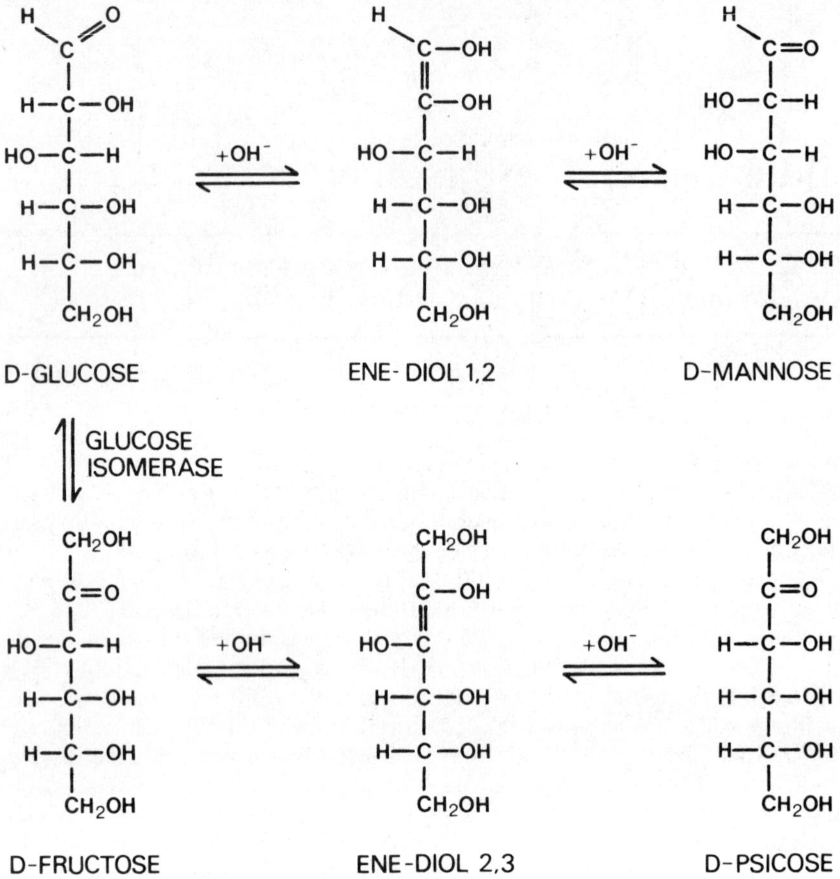

D-GLUCOSE ⇌ ($+OH^-$) ENE-DIOL 1,2 ⇌ ($+OH^-$) D-MANNOSE

GLUCOSE ISOMERASE (D-GLUCOSE ⇌ D-FRUCTOSE)

D-FRUCTOSE ⇌ ($+OH^-$) ENE-DIOL 2,3 ⇌ ($+OH^-$) D-PSICOSE

FIG. 7.1 Isomerizations of glucose and fructose.

by using an enzyme, would eliminate some of the problems, assuming that the enzyme would operate at an economically high temperature, at a sufficiently low pH and rapidly enough to minimize the chemical production of undesirable by-products.

To the classically-trained biochemist, glucose isomerase is an unexpected enzyme. The conversion of glucose to fructose in the majority of organisms investigated requires the isomerization of the phosphorylated sugar by the enzyme phosphoglucose isomerase (EC 5.3.1.9) so the discovery that an enzyme from *Pseudomonas hydrophila* which would isomerize xylose to xylulose could also convert glucose to fructose (Marshall and Kooi, 1957) was novel and exciting. Its affinity for glucose was 160 times smaller than that for xylose but this was sufficient for the enzyme to be commercially attractive. A patent based on this work was granted in 1960 (Marshall, 1960).

For the next decade work was apparently largely confined to Japan; more recently research into the production and use of this enzyme has been almost universal. Since then, various teams have worked simultaneously on glucose-isomerizing enzymes from different organisms. I shall trace the development of these enzymes, some of which have become industrially significant.

7.2.2 Glucose Isomerase Requiring Arsenate

The necessity to include arsenate in the reaction mixture for glucose isomerization meant that Marshall and Kooi's (1957) enzyme could not be exploited commercially. The same was true of the next few enzymes to be discovered. Tsumura and Sato (1960, 1961) demonstrated that extracts of the cells of the soil bacterium *Aerobacter cloacae* grown in a synthetic medium containing xylose contained an enzyme with properties similar to those of the *Pseudomonas hydrophila* enzyme (Marshall and Kooi, 1957), having a pH optimum of 8·0 and an optimum temperature of 42–43 °C. The next published information concerned the GI produced by *Aerobacter aerogenes*, Strain HN-56. This enzyme was produced when the organism was grown in media containing xylose, mannose or mannitol, the last being the most effective (Natake and Yoshimura, 1963). Because GI activity, but not xylose isomerase activity, could be detected in extracts of the cells grown in media containing mannose or mannitol, the authors concluded that this GI was not a xylose isomerase. It also required arsenate for the accumulation of fructose. The GI from *Escherichia freundii* HN-70 had similar properties (Natake and Yoshimura, 1963).

Natake and Yoshimura (1964a) transferred their attentions to *Escherichia intermedia* HN-500, which produced GI when grown on a very wide range of carbon sources but produced xylose isomerase only when grown in the presence of xylose. Once again it was essential to include arsenate in the reaction mixture if fructose was to be accumulated. The enzyme had similar properties to the previously-known GIs (Natake and Yoshimura, 1964b). The same workers (Natake and Yoshimura, 1964b) clarified the role of arsenate, which might have been expected to act by inhibiting reactions which would compete with GI for glucose or which would metabolize the fructose as it was formed. In fact the arsenate was essential for the activity of the enzyme itself. Eventually, Natake (1966) purified the glucose-isomerizing enzyme to near-homogeneity and revealed that it could isomerize glucose-6-phosphate and fructose-6-phosphate in the absence of arsenate and was probably identical with phosphoglucose isomerase (EC 5.3.1.9). Finally, Natake (1968) showed that the ratio of glucose-6-phosphate isomerization to arsenate-dependent glucose isomerization remained constant as the enzyme was purified. Kinetic experiments showed that a glucose–arsenate complex was formed and this was isomerized thus:

$$\underset{\text{G}}{\text{Glucose}} + \underset{\text{As}}{\text{Arsenate}} \rightleftharpoons \underset{\text{GAs}}{\text{Glucose-arsenate}}$$

$$\text{GAs} + \underset{\text{E}}{\text{Enzyme}} \rightleftharpoons \text{GAsE}$$

$$\text{GAsE} \rightleftharpoons \text{Fructose} + \text{As} + \text{E}.$$

7.2.3 Glucose Isomerases from *Lactobacillus* Species

Yamanaka first discovered that heterolactic acid bacteria could produce

Table 7.1 Properties of the 'Glucose Isomerase' of *Escherichia intermedia* HN-500. [*Natake (1968).*]

Substrate/Property	*Glucose (arsenate-dependent)*	*Glucose-6-phosphate*
Optimum temperature	50°	45°
Optimum pH	7·0	7·0
Thermal stability	Stable at pH 7·0–7·75	Stable at pH 7·0–7·75
Reaction equilibrium	G/F = 55/45	G-6-P/F-6-P = 60/40
Activation energy	15,300 cal/mol	8,300 cal/mol
Michaelis constant	1·6 M	$1{\cdot}4 \times 10^{-3}$ M
Maximum reaction rate	Vm/e 0·06 μmole fructose	1·0 μmole F-6-P

glucose isomerase. He had built up considerable expertise in the study of pentose isomerases in these bacteria (Yamanaka and Higashihara, 1962; Yamanaka, 1962a, b) without detecting any glucose-isomerizing ability associated with xylose or arabinose isomerases. It is not clear how GI activity was missed in Yamanaka's early work (1962a), because glucose was tested as a substrate for xylose isomerase from *Lactobacillus brevis* cells grown in a medium containing xylose in conditions used later to produce glucose isomerase from the same species (Yamanaka, 1963a). *Lactobacillus brevis* produced the highest yield of GI when compared with *Lactobacillus pentoaceticus, L. fermenti, L. mannitopoeus, L. gayonii, L. lycopersici, L. buchneri* and *Leuconostoc mesenteroides.* GI was produced in significant quantities only when the bacteria were cultured in media containing D-xylose; manganese ions were also essential (Yamanaka, 1963a). The enzyme was clearly different from the other 'glucose isomerases' known at that time as it did not require arsenate for activity and its activity was stimulated by manganese and cobalt ions.

The enzyme from *Lactobacillus brevis* was partially purified by treatment with $MnCl_2$, fractionation using ammonium sulphate, heat treatment and chromatography on DEAE-Sephadex (Yamanaka, 1963b). The ratio of glucose:xylose isomerase activity remained constant throughout the purification process, suggesting that the glucose and xylose isomerases were identical. The Michaelis constants for glucose and xylose were 0·52 M and $1{\cdot}3 \times 10^{-2}$ M respectively at the optimum pH for both activities of 6·0–7·0. There was no evidence for polyol dehydrogenase activity in the enzyme preparations, which indicated that the isomerization was direct and did not involve the participation of sugar alcohols. However, sorbitol and xylitol inhibited both glucose and xylose isomerization.

Yamanaka (1968) purified and crystallized the xylose isomerase from *Lactobacillus brevis,* using techniques similar to those employed earlier. The pure enzyme showed ribose isomerase activity in addition to xylose and glucose isomerase activity; the ratio of activities (at saturating substrate concentrations) was 100:108:16·5 respectively for xylose, glucose and ribose. Manganese ions were essential for xylose isomerase activity, and both manganese and cobalt ions were included in the assay mixtures for glucose and ribose isomerase. No metal-ion specificity studies were undertaken. The molecular weight of the enzyme was found to be 195,000. A patent based on this work was granted in Japan in 1965 (Yamanaka, 1965).

Although this enzyme appears to have attractive properties for commercial development, in particular its low pH optimum (Yamanaka, 1963a, 1968), it is less stable at high temperatures than its competitors. Consequently it has not been employed in a commercial process. Kent and Emery (1973) studied the enzyme from *Lactobacillus brevis* in some detail and demonstrated that autolysis at 40 °C and pH 7·0 was much the most effective method of those tested for the release of the enzyme from washed cells. The fructose contents of reaction mixtures at equilibrium were high, 58, 64 and 63% (of total sugars) respectively at 35 °, 50 ° and 60 °C.

The production of the GI of *Lactobacillus brevis* and its immobilization on DEAE cellulose have been patented by Baxter Laboratories Inc. (1972). The enzyme has also been immobilized on to cellulose using titanium chloride (Emery *et al.*, 1974).

7.2.4 Glucose Isomerase from *Streptomyces* Species

Streptomyces species have probably been studied more intensively as sources of GI than those of any other genus. The ability of extracts of *Streptomyces phaeochromogenes* strain SK to isomerize glucose when the cells were grown in a medium containing xylose was noted by Tsumura and Sato (1965). The enzyme was purified by sonication of the cells, fractionation of the extract using ammonium sulphate and heat treatment. It isomerized only glucose and xylose and required magnesium ions for full activity, the K_m for magnesium ions being $2{\cdot}3 \times 10^{-3}$ M. Cobaltous ions activated the GI in the presence of magnesium ions and, at 10^{-3} M and higher, protected the enzyme against heat denaturation so that it lost very little activity during 10 min at 80 °C. The temperature optimum for GI activity was high, 80 °C in the absence of cobalt ions, 90 °C in the presence of 10^{-3} M cobaltous ions. The glucose substrate also appeared to protect the enzyme against heat damage. At 60 °C the equilibrium level of fructose was 52% and the K_m for glucose was 0·30 M. All these properties were very attractive commercially but unfortunately the pH optimum was high, pH 9·3–9·5. The search was then on for even more attractive strains of *Streptomyces*. Tsumura *et al.* (1967) persisted with *S. phaeochromogenes* strain SK and showed that if cobaltous ions at 10^{-3} M were included in the fermentation medium the pH range for enzyme activity was extended so that the activity at pH 7·5 was similar to that at pH 9·0.

Takasaki (1966) searched for organisms which were capable of assimilating xylan and producing GI and found a *Streptomyces* species of the *cinereus* series which he designated Strain YT-No 5. (Later this was identified as *S. albus* (Takasaki *et al.*, 1969a.) Xylose or xylan were essential for the production of GI and yields of enzyme were improved by the inclusion of cobaltous ions (10^{-3} M) in the fermentation medium. The temperature optimum of the enzyme was 70 °C, lower than that of the GI from *S. phaeochromogenes* (Tsumura and Sato, 1965), but the pH optimum was 7·0. The requirements for magnesium and cobalt ions were similar to those of the GI from *S. phaeochromogenes*. At equilibrium at 60 °C reaction mixtures contained 54% fructose. The K_m for glucose and fructose was $1{\cdot}4 \times 10^{-1}$ and $2{\cdot}3 \times 10^{-1}$ M respectively (Takasaki, 1967). Later, the enzyme was crystallized and studied in detail (Takasaki *et al.*, 1969b). Its molecular weight was estimated to be

157,000, the pH optimum at 70 °C was 8·0–8·5 and the temperature optimum was now 80 °C. Unlike the enzyme from *Lactobacillus brevis* (Yamanaka, 1968), this GI did not isomerize ribose. The pH stability was wide, no activity being lost during incubation for 3 h at 20 °C at any pH above 4·0 and below 11·0. Maximum activity was obtained with 10^{-3} M cobaltous ions and 0·1–0·15 M magnesium ions. The crystalline enzyme contained 1·4 atoms of cobalt and 0·3 atom of magnesium per molecule of the enzyme.

Further information of GIs from *Streptomyces* species has to be gleaned from the patent literature. Takasaki's work was patented in the United Kingdom in 1968 (Agency of Industrial Science and Technology, 1968). This patent describes the production of GI by *Streptomyces flavovirens, S. echinatus, S. achromogenus* and *S. albus* strains YT-4 and YT-5. The latter two were by far the most efficient producers of the enzyme (Cotter *et al.* (1971); Standard Brands Inc. (1972)) patented the use of *Streptomyces* species ATCC 21175 and ATCC 21176, which produces GIs with properties similar to those of the other *Streptomyces* species (Takasaki *et al.*, 1969b). Strain ATCC 21175 was identified as *S. wedmorensis* (Dworschack and Lamm, 1972): this was the source of the GI used to produce 500,000 tons of high-fructose corn syrup by 1974 (Schnyder, 1974).

Meanwhile, Japanese workers studied the species *Streptomyces venezuelae* and *S. olivochromogenes*, in particular *S. olivochromogenes* ATC 21114 (Iisuka *et al.*, 1972) which is apparently unable to utilize xylan. Miles Laboratories Inc. (1972) produced GI using *Streptomyces olivaceus* NRRL 3588. This enzyme operated at pH 8·5 and at 60–70 °C and required magnesium and cobalt ions. A mutant of this organism, NRRL 3916, is the source of the GI used on an industrial scale by Miles-Cargill.

Weber (1975) isolated a strain of *Streptomyces glaucescens*, ETC 22794 which is distinguished by its ability to produce comparatively large amounts of extracellular GI. Very recently, CPC International Inc. (1975a, b, c) have described the production of a series of mutants of *S. olivochromogenes* ATCC 21114 which are capable of producing increased yields of GI when grown in media containing neither xylose nor cobalt ions. It is reasonable to suppose that one of these strains is the source of the GI used in the production of Corn Products' high-fructose corn syrup ('Invertose') (Godzicki, 1975).

7.2.5 Glucose Isomerases from *Bacillus* Species

A screen of over 300 organisms, grown in a medium containing xylose at 45 °C, resulted in the selection for extensive study of a strain (HN-68) of *Bacillus coagulans* (Yoshimura *et al.*, 1966). This organism produced GI only when grown in the presence of xylose, and the yield of enzyme was improved by the inclusion of manganese ions in the growth medium. The properties of the GI were encouraging for industrial development: the pH optimum was 7·0 and the optimum temperature for fructose production was 60 °C. Cobaltous ions were more effective than magnesium ions in promoting GI activity. Purification of the enzyme (Danno *et al.*, 1967) by treatment of a cell-free extract with $MnSO_4$ solution, ammonium sulphate fractionation and chromatography on DEAE-Sephadex gave a preparation which was homogeneous on polyacrylamide gel electrophoresis. This treatment apparently increased the temperature optimum to 75 °C. The molecular weight, determined after crystallization of the enzyme, was 175,000 (by sedimentation) or 160,000 (by gel filtration) (Danno, 1970a). The iso-electric point was pH 4·9. Like the GI from *Lactobacillus brevis* (Yamanaka, 1968), the enzyme from *Bacillus coagulans* required cobalt ions to isomerize glucose and ribose but manganese ions to isomerize xylose (Danno, 1970b). The K_m values were 9×10^{-2} M, $1{\cdot}1 \times 10^{-3}$ M and $7{\cdot}7 \times 10^{-2}$ M respectively for glucose, xylose and ribose and the relative V_{max} values at 40 °C were 0.52, 1·1 and 0·25 mg/min. The kinetic constants and induction studies indicated that the 'glucose isomerase' was truly xylose isomerase. The thermal stability was as good as that of the GI of *Streptomyces albus* (Takasaki *et al.*, 1969b) but the enzyme was less stable at high pH values. Cobalt, manganese and nickel ions all protected the enzyme against heat inactivation. Tris buffer, xylitol, sorbitol and mannitol were all competitive inhibitors of all three isomerase activities, which implied that glucose, xylose and ribose were all isomerized at the same active centre. Danno (1971) studied the roles of the metal ions in the isomerizing reactions in some detail. He concluded that four cobalt ions were bound per enzyme molecule and that the active form of the enzyme was that which contained the cobalt (or manganese) ions. Activation of the enzyme by the metal ions was highest at pH 6·0 and decreased rapidly with increasing pH. Since the binding of the substrate to the enzyme was independent of the presence of metal ions it seemed unlikely that the cobalt or manganese ions catalysed the reaction. The cobalt and manganese ions were bound at different sites and Danno (1971) proposed an elegant scheme for the roles of the two metal ions (Figure 7.2).

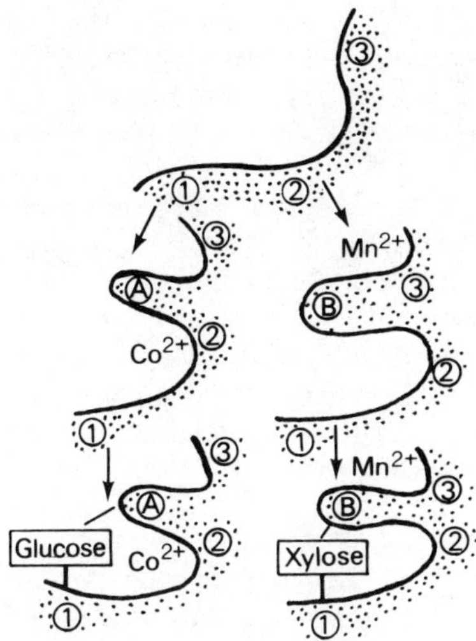

FIG. 7.2 Proposed mechanism for the activation of the xylose (glucose) isomerase of *Bacillus coagulans* by cobalt ions (for glucose isomerization) and manganese ions (for xylose isomerization). [*Danno, 1971.*

The GI from a strain of *Bacillus coagulans* is offered for sale by Novo Industri A/S as 'Sweetzyme' (Zittan *et al.*, 1975).

7.2.6 Other Glucose Isomerases

Information on GIs produced by micro-organisms not previously mentioned, some of them of great industrial significance, appears almost exclusively in the patent literature. The most important is probably that produced by *Arthrobacter* species designated NRRL-B-3724, 3725, 3726, 3727 and 3728 (Lee *et al.*, 1972; R. J. Reynolds Tobacco Co., 1973). The last three strains produced GI when cultured in the absence of xylose or xylan and gave yields of enzyme similar to those produced by strains B-3724 and 3725 cultured in media containing xylose or xylan. The pH optimum of the enzyme was 8·0 and the operable temperature range was 50–90 °C. An important feature is that this enzyme does not require cobalt ions for activity or stability. To the best of my knowledge no further information is available regarding the properties of this enzyme. Since some of the *Arthrobacter* strains require xylose to be able to produce GI it is probable that this enzyme is also truly a xylose isomerase.

Anheuser-Busch studied various micro-organisms as sources of GI and patented the use of the enzyme from *Aerobacter levanicum* (Shieh *et al.*, 1974a). This GI was produced only when the fermentation medium contained xylose (from birchwood sulphite liquor) and cobalt ions. Although the enzyme could be released from the bacterial cells by sonication, only whole, air-dried cells were used to isomerize glucose. The pH used was between 6·5 and 7·5 and the temperature was maintained between 50 and 60 °C. Cobalt and magnesium ions were both used at 0·6 mM. Although there are no obvious drawbacks to the use of this enzyme, Anheuser-Busch diverted their attentions to the GI produced by *Actinoplanes* species. These species are capable of producing the enzyme in the absence of xylose or xylan (Scallet *et al.*, 1974; Anheuser-Busch Inc., 1975) and in media containing relatively cheap sources of nitrogen. All the *Actinoplanes* species tested produced GI but *Actinoplanes missouriensis* NRRL-B-3342 was very much the best. The pH optimum of the enzyme was 7·0 and it is apparently the most stable of all the known GIs since it retained activity at 90 °C (though the maximum time tested was only 25 mins). Both cobalt and magnesium ions were required for high activities: the magnesium was essential for activity while cobalt enhanced the activity. The optimum concentrations of magnesium and cobalt ions were 3·0 mM and 0·3 mM respectively. This was another xylose isomerase, capable also of isomerizing ribose, the relative activities being to glucose 100, xylose 49 and ribose 16 (Scallet *et al.*, 1974). This GI is used commercially by Anheuser-Busch Inc.

Standard Brands Inc. (1975) have extended their studies to include micro-organisms of the genera *Nocardia*, *Micromonospora*, *Microbispora* and *Microellobospora*. Each of these organisms produced GI when supplied with xylose and cobalt ions; the enzymes were active at pH 6·7 and 70 °C in the presence of 20 mM $MgSO_4$. These organisms show no obvious superiority to better-known producers of GI.

Table 7.2 Yields of Glucose Isomerase Produced by the Various Organisms Tested.

Organism	*Yield (GIU/litre of fermentation medium)*	*Assay temp.*	*Assay* pH	*Reference*
Streptomyces sp	20–1,160	70	7·5	Agency (1968)
Streptomyces ATCC 21114	200	60	7·5	Iisuka *et al.* (1972)
Streptomyces phaeochromogenes	24	70	7·2	Baxter Labs (1972)
Streptomyces wedmorensis	2,200–6,800	70	7·2	Dworschack *et al.* (1972)
Streptomyces olivochromogenes	60	60	7·5	CPC Internat. (1972)
Arthrobacter NRRL-B-3724	3,340			
NRRL-B-3726	4,720	60	7·5	Reynolds (1973)
NRRL-B-3727	2,220			
NRRL-B-3728	4,440			
Streptomyces wedmorensis ATCC 21230	560–2,500	60	7·2	Agency (1974)
Streptomyces ATCC 21175	4,640/g			
Mutant 1	7,540/g	70	7·5	Bengtson & Lamm (1974)
Mutant 2	6,680/g			
Mutant 3	6,000/g			
Streptomyces olivaceus NRRL-3583	2,560	60	7·8	Miles Labs (1974)
Mutant NRRL-3916	2,960			
Actinoplanes missouriensis	2,500–35,200	75	7·0	Anheuser-Busch (1975)
Nocardia asteroides	400			
Nocardia dassonvillei	400			
Micromonospora coerula	320	70	6·7	Standard Brands (1975)
Microbispora rosea	160			
Microellobospora flavea	160			
Streptomyces glaucescens ETH-22794	120–840	70	7·0	Weber (1975)
Streptomyces olivochromogenes Mutant CPC 3	4,800–11,440			
CPC 4	5,700–9,680	60	7·5	CPC Internat. (1975)
CPC 8	3,960–4,440			

(1 GIU = amount of enzyme which will convert 1 μmol glucose to fructose/min in the conditions of the assay.)

7.3 FERMENTATION ADVANCES IN THE PRODUCTION OF GLUCOSE ISOMERASE

The objectives of fermentation research have been:

(1) To increase yields of GI.
(2) To replace fermentation media containing xylose by cheaper media.
(3) To eliminate the need for cobalt ions in the fermentation media.

7.3.1 Increase in Yields

Table 7.2 summarizes and attempts to compare the GI activities which have been obtained from the various organisms which have been tested. It is clear from this that some workers have been fortunate in obtaining active strains of organisms while others are struggling with comparatively poor organisms. The 'record' is apparently held by Anheuser-Busch's (1975) *Actinoplanes missouriensis.* Anheuser-Busch (1975) describe the importance of using the best available nitrogen source in order to achieve high yields of GI (Table 7.3). It was essential that all sludge was removed from the corn steep liquor

Table 7.3 Effect of Organic Nitrogen Source on the Production of Glucose Isomerase by *Actinoplanes missouriensis*. [*Anheuser-Busch (1975)*.]

Organic nitrogen source	*Manufacturer*	*Activity (GIU/ml culture)*	%
Corn steep liquor	Anheuser-Busch	17·8	100
O.M. peptone	Amber Lab.	13·4	75
Casein hydrolysate (Amber EHC)	Amber Lab.	13·0	73
Bacto-soytone	Difco Lab.	11·8	66
Yeast extract (BYF-100)	Amber Lab.	10·3	58
Distiller's dried soubles	National Distillers	7·9	44
Yeast extract (BYF-300)	Amber Lab.	5·8	33
Bactopeptone	Difco Lab.	5·2	29
Yeast extract (BYF 50X)	Amber Lab.	4·3	24
Malt extract	Difco Lab.	2·5	14
Atlantic Menhaden peptone	Haynie Products	1·1	6

and that this was neutralized before use, preferably using sodium or ammonium hydroxide. There is no evidence in the literature that such a systematic investigation of nitrogen sources has been done in the study of GI production by other organisms.

Some large improvements in yield have been achieved by mutating active organisms. Bengtson and Lamm (1974) used ethyleneimine to produce a mutant strain of *Streptomyces* ATCC 21175 and obtained a 62% increase in enzyme yield. Miles Laboratories (1974a) obtained a 16% increase in yield by mutating *Streptomyces olivaceus* NRRL 3583 using UV light. Recently, CPC International Inc. have described (1975a, b, c) the mutation of *Streptomyces olivochromogenes* ATCC 21114 by irradiating spores with UV light to produce a series of strains, some of which were capable of producing 50% more GI than the parent strain. It is unfortunate that no information is yet available on the production by Novo Industri A/S of the GI from *Bacillus coagulans* other than that 'a specially selected strain' is employed (Zittan *et al.*, 1975).

It seems probable that further screening of unknown, probably thermophilic, organisms will detect more satisfactory sources of GI.

7.3.2 Replacement of Xylose

It was rapidly recognized that pure D-xylose was too rare and expensive to be used in a commercial fermentation process. Takasaki (1966) introduced the use of *Streptomyces* species which were capable of utilizing xylan in the form of materials such as wheat bran, corn hulls and corn-cob. This work was patented in the U.K. (Agency of Industrial Science and Technology, 1968), the favoured organism being *Streptomyces albus* strain YT-4, which produced satisfactory amounts of GI when cultivated in media containing 4% corn steep liquor, 0·1% $MgSO_4 \cdot 7H_20$, 0·3% NaCl with material containing 0·5% xylan. Cotter *et al.* (1971) (Standard Brands Inc. 1972) used corn-cob hydrolysate to supply 1% xylose in a medium also containing 1% sorbitol, 0·75% glucose, 4% steep water and 0·024% cobaltous ion. Isuka *et al.* (1972), using *Streptomyces olivochromogenes* and *S. venezuelae*, employed pure xylose in fermentation media and demonstrated that higher yields of GI were obtained when 0·5% xylose and 0·5% starch were used instead of 1·0% xylose. Dworschack *et al.* (1972) described the use of various sources of xylose, both free and combined, in the production of GI by *Streptomyces* ATCC-21175 and 21176.

Workers at the Agency of Industrial Science and Technology (1974) appreciated that the use of 1–5% of D-xylose (free or combined) as an inducer of GI production impeded the growth of the micro-organisms and also that appreciable amounts of the xylose remained in the culture medium at the end of the fermentation. They demonstrated that lower levels of xylose, xylobiose or xylose plus xylobiose could be used (Figures 7.3 and 7.4) and

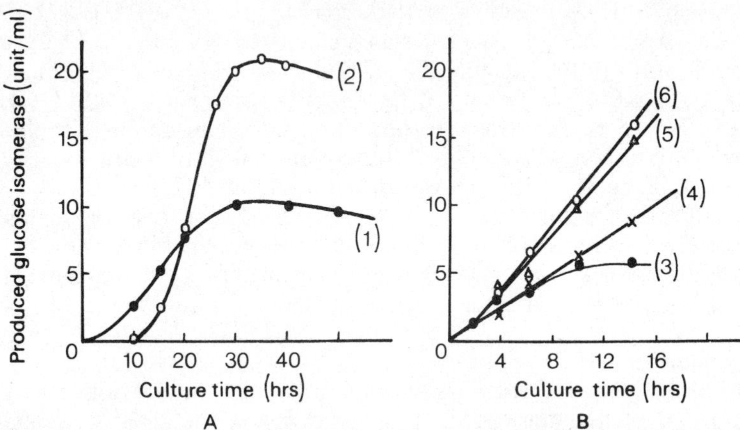

FIG. 7.3 A. Time-course of (1) cell growth of *Streptomyces albus* and (2) production of glucose isomerase. B, Time-course of glucose isomerase production in the presence of (3) 0·5% xylose, (4) 0·5% xylobiose, (5) 0·4% xylose + 0·1% xylobiose, (6) 0·3% xylose + 0·2% xylobiose. [*Agency, 1974*

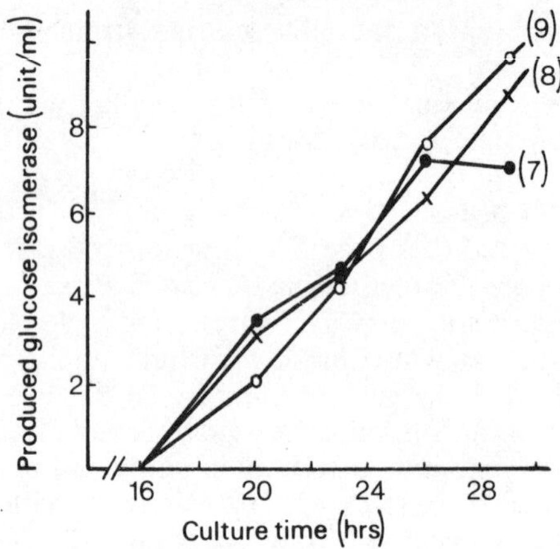

Fig. 7.4 Time-course of glucose isomerase production by *Streptomyces wedmorensis*. (7) 0·5% xylose added at the 16th hour of culture, (8) 0·01% xylobiose added at hourly intervals, (9) 0·025% xylose added at hourly intervals. [*Agency, 1974*.

that similar amounts of GI were produced when the xylose was present at inoculation, when it was added in one dose after 16 h and when it was added in small amounts at hourly intervals. The yield of GI was improved by growing *Streptomyces albus* ATCC 21132 in an initial medium containing low levels of a xylose source together with corn steep liquor for 30 h, then adding small amounts of xylose at hourly intervals. Xylan and xylan hydrolysate were equally effective. The highest activities of all were achieved by adding xylose continuously after the 28th hour of culture. The CPC International (1975a) mutants appear to be the first *Streptomyces* strains capable of producing GI in the absence of xylose or xylose-containing material.

The xylose problem has been overcome, or more accurately does not occur, when GI is produced using *Arthrobacter* species NRRL B-3726, 3727 and 3728 (Lee *et al.*, 1972; R. J. Reynolds Tobacco Co., 1973). The growth medium for GI production contained 2% glucose, 0·6% meat protein, 0·15% yeast extract, 0·6% $(NH_4)_2SO_4$, 0·2% KH_2PO_4 and 0·01% $MgSO_4 \cdot 7H_2O$. Similarly *Actinoplanes missouriensis* produces GI when supplied simply with 3% corn steep liquor, 0·1 mM cobaltous ions and 0·05 mM cupric ions (Scallet *et al.*, 1974; Shieh *et al.*, 1974b; Anheuser-Busch Inc., 1975).

Although the original strains of *Bacillus coagulans* required xylose to produce GI (Yoshimura *et al.*, 1966) it is believed that Novo Industri A/S have developed a strain which is capable of producing the enzyme when grown in the absence of free or combined xylose.

7.3.3 Elimination of Cobalt

This appears to have received less attention than the two previous topics but is important in that the cobalt ions remaining in the spent fermentation medium may constitute a serious environmental hazard. The CPC International (1975b) mutants do not require cobalt in order to produce GI, nor do any of the *Arthrobacter* species studied by R. J. Reynolds Tobacco Co. (1973).

7.4 IMMOBILIZATION OF GLUCOSE ISOMERASE

Glucose isomerase is an expensive enzyme which has come into commercial use since the potential advances to be gained by the use of immobilized enzymes have been widely appreciated. It is also readily available, uses a low-molecular weight substrate and is rapidly and easily assayed by conventional techniques. Consequently it is a very convenient model enzyme for use in immobilization studies and the patent literature includes details of the immobilization of GI by each of the many methods available. Inevitably, not all of these methods are applicable economically: it appears to be the straightforward methods which have found industrial applications.

7.4.1 'Fixing' of Whole Cells Containing Glucose Isomerase

Almost every known GI is an intracellular enzyme and in many cases whole microbial cells have been used to isomerize glucose (e.g. Yoshimura *et al.*, 1966; Tsumura *et al.*, 1967; Takasaki, 1969a). However, cells of *Streptomyces albus* autolysed at temperatures below 50 °C (Takasaki *et al.*, 1969a) but retained their GI if they were heated at temperatures over 60 °C for 10 min. before use at lower temperatures. Tsumura *et al.* (1967) dehydrated cells of *Streptomyces phaeochromogenes* by various methods, all of which failed to fix the GI within the cells, presumably because insufficiently high temperatures were employed. They retained 90% of the activity of resting cells by spray drying using a temperature of 100 °C but unfortunately did not describe the properties of this preparation. GI immobilized within cells as described by Takasaki *et al.* (1969a) was used in the first industrial process to produce high-fructose glucose syrup by the Clinton Corn Processing Company (Schnyder, 1974).

Anheuser-Busch Inc. apparently use spray-dried cells of *Actinoplanes missouriensis* in an industrial process; it has not been revealed whether such cells may be re-used (Scallet *et al.*, 1974).

It is remarkably easy to immobilize GI within the cells of *Streptomyces* species since this has been achieved merely by incubating the cells in aqueous solutions of inorganic salts (Lamm *et al.*, 1974). A wide range of salts was used at concentrations between 0·01 and 0·5 M to immobilize up to 100% of the isomerase. Cobaltous chloride was the most-favoured salt. No indication was given of the re-usability of GI immobilized in this manner.

Heady and Jacaway (1974) produced another very straightforward method of immobilizing GI within cells of *Streptomyces* species for use in a continuous

Table 7.4 Comparison of Batch and Column Isomerization of a 95 DE Starch Hydrolysate. [*Heady and Jacaway (1974a).*]

Batch, 50 hours, pH 6·25	*Feed*	*Product*	*Difference*
Ash, % d.b.	1·0	1·8	0·8
Organic acids, % d.b.	0·05	0·08	0·03
Colour	4	200	196
Column, 1–2 hours, pH 8			
Ash, % d.b.	0·5	0·6	0·1
Organic acids, % d.b.	0·03	0·04	0·01
Colour	4	11	7

process. This involved adding two parts (by weight) of $Mg(OH)_2$ for each part of cells in fermentation broths and filtering to produce a cake which was dried at room temperature. This preparation retained a constant activity for 5 days and was used in all for 10 days, during the last two of which the temperature was raised from 60 to 70 °C. The $Mg(OH)_2$ could be substituted by diatomaceous earth filter aid. Preparations such as these showed many advantages over enzymes or cells used in batch processes (Table 7.4).

Lloyd *et al.* (1974) patented a similar method in which cells of *Arthrobacter* species were simply trapped in filter aid within a leaf filter and used continuously, with only a small loss of activity, for 15 days.

7.4.2 Physical Entrapment of Whole Cells

Various methods have been described, usually in patents, for the entrapment or encapsulation in polymeric materials of whole cells retaining GI activity. Some of these have assumed commercial importance.

Nielsen and Markussen (1974a, b) developed a process using a machine described as a 'Marumeriser' to produce uniformly-sized solid spheres containing protein and binding agents of various types. Although it has not been announced that this is so, it seems that this is basically the process used by Novo Industri A/S to immobilize the GI from *Bacillus coagulans* as 'Sweetzymes' (Zittan *et al.*, 1975).

R. J. Reynolds Tobacco Co. (1974) have patented a process similar in concept to the Novo process in which *Arthrobacter* cells containing GI are formed into an aggregate by using polyelectrolyte flocculating agents and, sometimes, filter aid. No information about the stability of the GI in this preparation is available.

Kolarik *et al.* (1974) immobilized whole cells (presumably of *Bacillus coagulans*) obtained from Novo Enzyme Corporation, in cellulose acetates. High ratios of cells : cellulose acetate were used, usually 1 : 2, and fibres and membranes were produced by extruding or casting emulsions formed by mixing aqueous slurries of the cells with solutions of the cellulose acetate in methylene chloride. The experiments were not too successful because cells leaked from the membranes; moreover the working life of the preparation was short and the activity of the preparation was limited by the rate of diffusion of substrate to the cells.

7.4.3 Chemical Entrapment of Whole Cells

The idea that active microbial cells may be entrapped in polymeric materials and fixed there chemically has been tested by several groups of workers. Vieth *et al.* (1973) immobilized cells of *Streptomyces phaeochromogenes* containing GI by first heat treating them as described by Takasaki *et al.* (1969), and then stirring them with a slurry of hide collagen, adjusting the pH to 11·2 and casting the mixture on a suitable surface. The resulting membrane was tanned using glutaraldehyde or formaldehyde. The resulting preparation was used continuously for 40 days at 70 °C in a plug-flow type of reactor. Miles Laboratories Inc. (1974b) simply treated cells of *Streptomyces olivaceus* NRRL-3583 with glutaraldehyde and showed that this treatment allowed them to be re-used nine times with only small losses of GI activity, in contrast to untreated cells which lost all their activity on being used twice. This preparation is to be used on a commercial scale by a combine of Miles Laboratory and Cargill (Eurofoods, 1975).

Moskowitz (1974) (Baxter Laboratories Inc., 1975) described a method for immobilizing whole cells using diazotized primary diamino compounds as cross-linking agents. Diazotized diamino compounds containing pyridine, pyrimidine or acridine moities were preferred. This method was used to cross-link whole cells of *Streptomyces phaeochromogenes* by stirring the cells with the diazotized reagent at 4 °C for up to 48 h. The yield of immobilized GI was usually greater than 50% of that in the fresh cells and the activity was retained unimpaired for the first 3 to 5 cycles of use. In one case 48% of the original activity remained after 20 cycles of batchwise operation. The original cells lost most of their activity after two cycles of use.

7.4.4 Immobilization of Soluble Glucose Isomerase

Although GI has been used successfully when fixed within whole microbial cells, the advantages gained by immobilizing soluble preparations of the enzyme on to solid supports have meant that such preparations are superseding some of the earlier processes. Enzymes immobilized on to supports selected to give excellent flow characteristics may be used satisfactorily in continuous processes with concomitant savings in capital equipment.

Baxter Laboratories Inc. (1972) immobilized the GIs from both *Streptomyces phaeochromogenes* and *Lactobacillus brevis.* The enzymes were released from the cells and the solutions clarified, then stirred with DEAE-cellulose at pH 8·5 in 0·02 M tris buffer. The enzyme-DEAE-cellulose complex was used after washing with 0·02 M tris pH 8·0. The properties of the immobilized enzyme are compared with those of the soluble enzyme in Table 7.5 and

Table 7.5 Comparison of Properties of Glucose Isomerase Free and Bound to DEAE-cellulose. [*Baxter Laboratories Inc. (1972).*]

Property	*GI*	*DEAE-GI*
pH optimum	8·5–10·0	7·5–8·5
pH stability	5·0–8·0	4·5–9·0
Temperature optimum	80–85°	75–85°
Temperature stability at 70°	20 hours	Stable
Metal iron requirement	Mg^{2+} (10^{-2} M) Co^{2+} (10^{-3} M)	Mg^{2+} (10^{-2} M) Co^{2+} (10^{-3} M)
Conversion of glucose to fructose	35–40%	45–50%
Loss of enzyme activity after each conversion	100%	See Fig. 7.6
Amount of enzyme needed to convert 100 g of glucose to fructose (40%) in 24 h, at 70°, pH 7·0	1,800 units	1,200 units

Figure 7.5. The DEAE-cellulose could be used to purify the GI if necessary and could be recovered for re-use simply by treatment of the exhausted complex with 0·6 M NaCl solution.

Thompson *et al.* (1974) describe a similar process in which GI was released from cells of *Streptomyces* ATCC 21175 by treatment with a cationic detergent and purified by heat treatment at 60 °C and pH 6·7–6·8, then stirred with

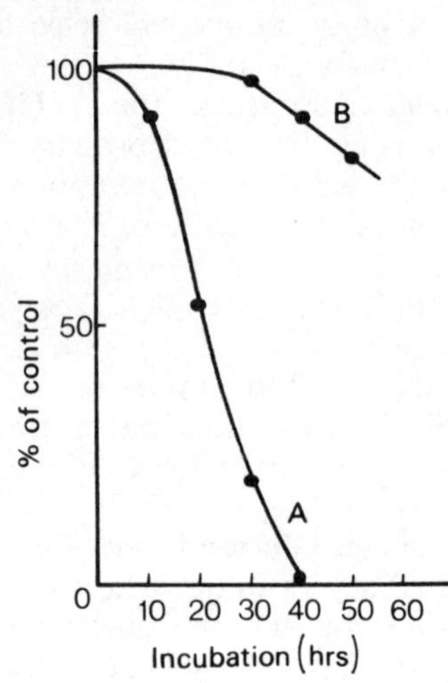

Fig. 7.5 Stability of glucose isomerase from *Streptomyces phaeochromogenes* at 70 °C. A, In solution. B, Immobilized on DEAE-cellulose. [*Baxter Laboratories, 1972.*

DEAE-cellulose to give a final preparation containing 35 GI units/g. of complex. This preparation had a half-life of 198 h. and could be operated at about 0·6 column volumes per hour in a column 25 cm high × 1 cm in diameter, giving an initial conversion of glucose to fructose of 49·6%. Amberlite IRA-938 could also be used; its complex with GI was more stable than the DEAE-cellulose-GI complex but its superior flow characteristics apparently could not be utilized. The *Streptomyces* GI immobilized on DEAE-cellulose is used in the Clinton Corn Processing Company's semi-continuous plant to produce high-fructose corn syrup ('Isomerose') (Schnyder, 1974).

Messing (1975) immobilized GI from a species of *Streptomyces* on porous alumina, the pores having diameters between 140 and 220Å, by repeatedly evaporating solutions of GI on to the support. This process immobilized 30%–75% of the enzyme applied and produced material with an activity of 260 GIU/g. (1 GIU converts 1 μmol of glucose per mm to fructose.) The maximum half-life was 49 days, the extent of the half-life depending on the inorganic ion content of the glucose substrate (Table 7.6). This preparation is suitable for continuous use in plug-flow reactors if the mesh size of the alumina particles is between 25 and 60 mesh. Zirconia, titania and alumina/titania could be used similarly (Messing, 1974). The use of GI immobilized on controlled pore alumina in the continuous production of fructose-enriched glucose syrups has recently been described (Messing and Filbert, 1975). An important observation was that once the immobilized enzyme had been equilibrated with cobalt ions the cobalt could be omitted from the substrate solution without loss of GI activity. The rights to this process have been acquired by CPC International Inc. and it will presumably be used in a high-fructose corn syrup plant scheduled to commence production in 1976 (Eurofood, 1975).

Emery *et al.* (1974) prepared an immobilized GI by linking the enzyme

Table 7.6 Effect of Inorganic Ion Content of Substrate on the Life of Glucose Isomerase Immobilized on Porous Alumina. [*Messing (1975)*.]

Example	*Enzyme loading* (GIU/g)	% *enzyme retained*	*Additions to substrate* (M)	*Initial* ml/h	$T_{\frac{1}{2}}$ (days)
1	573	76·5	0·004 $MgSO_4$ 0·001 $CoCl_2^4$ 0·004 Na_2SO_3	122	49
2	563	70·8	0·005 $MgCl_2$ 0·001 $CoCl_2$ 0·004 NH_4HCO_3	112	23
3	206	36·5	0·005 $MgCl_2$ 0·001 $CoCl_2$ 0·004 Na_2SO_3	45	47
4	192	32·3	0·005 $MgCl_2$ 0·001 $CoCl_2$ 0·004 NH_4HCO_3	38	17
5	204	39·9	0·005 $MgCl_2$ 0·004 NH_4HCO_3 NoCo^{2+}	48	18
6	407	49·2	0·005 $MgSO_4$	74	38

from *Lactobacillus brevis* to cellulose treated with titanic chloride. Approximately 40 GIU were linked per gram of solid. These studies were extended by Kent and Emery (1974) who showed that the amount of activity linked to the cellulose was approximately constant between pH 5·0 and pH 9·0. Immobilization decreased the pH optimum from 7·0 to 6·0 but decreased the temperature optimum from 60° to 50 °C. It was concluded that the coupling method was successful but that the low density of the immobilized enzyme preparation would not allow the use of this preparation in a continuous reactor.

Monsanto Co. (1974) immobilized GI from an undisclosed source on large-pore polyethylene discs by permeating the discs with a solution of polyacrylonitrile in dimethylsulphoxide, washing with water to precipitate the polyacrylonitrile, washing and cycling the enzyme through the discs. Finally the enzyme was fixed by cycling 1% (v/v) glutaraldehyde through the preparation. The activity of the immobilized GI was not disclosed.

Dinelli *et al.* (1974) described an elegant procedure by which enzymes were entrapped in a filament of cellulose acetate. GI from a *Streptomyces* species was purified by acetone precipitation, dissolved in aqueous glycerol (72:25 v/v) and made into an emulsion with a 7% (w/v) solution of cellulose triacetate in methylene dichloride. The emulsion was spun through a spinneret into toluene and the resulting filaments dried *in vacuo*. The total activity immobilized was not disclosed; the fibres retained some activity for 45 days but the activity dropped steadily throughout this period. A similar procedure was used to immobilize GI and AG together in fibres which could convert starch previously liquefied using α-amylase to a mixture of glucose, fructose and oligosaccharides.

Research Corporation (1975) have prepared membranes of protein containing GI by swelling collagen film at pH 3, washing and soaking it in a solution of the enzyme for 16 h before drying in air at room temperature. This produced a stable re-usable GI preparation.

7.5 THE SUBSTRATE: QUALITY OF THE GLUCOSE FEEDSTUFF

Glucose isomerase is completely specific for monomeric D-glucose. The maltose, maltotriose and higher maltooligosaccharides present in glucose syrups remain untouched by the enzyme. Consequently the final amount of fructose which may be produced in a syrup is dependent on the amount of glucose present in the substrate. An accepted composition of high-fructose glucose syrups in commerce is:

Fructose 42%
Glucose 50%
Maltose 6%
Maltotriose 2% (e.g. Schnyder 1974)

which implies that the starting syrup had a DE of 95–96. Such DE levels are not easily achieved using acid hydrolysis of starch followed by saccharification by amyloglucosidase; and indeed enzyme manufacturers claim that it is impossible. However, since the Clinton Corn Processing Co., which has the longest experience of producing fructose-enriched syrups, uses the acid-enzyme process (Hamilton *et al.*, 1974) there must be no great objection to using glucose syrup produced in this way.

Materials that might be expected to interfere with the action of GI are sugars or their derivatives, which might act as competitive inhibitors, and metal ions. In practice it appears that none of the many disaccharides and oligosaccharides which may occur in small amounts in 90^+ DE glucose syrups (Birch and Kheiri, 1970) interferes with GI activity (at the concentrations present) and thus it is possible to isomerize the glucose present in syrups of any DE. Clinton produced a syrup ('Isomerose 30') starting with 75 DE glucose syrup and ending with

Glucose 43%
Fructose 14%
Maltose 31%
Higher saccharides 12%

and recently an isomerized 63 DE syrup has found a market in the soft drinks industry (Felton, 1975). The Agency of Industrial Science and Technology (1968) isomerized the 'hydrol' obtained in the crystallization of glucose from starch hydrolysate to produce a syrup containing

Fructose 34·7%
Glucose 42·4%
Oligosaccharides 22·9%

This is perhaps the best example in the literature demonstrating that GI can act in comparatively impure glucose syrups. Metal ions, especially calcium, are important inhibitors of GI. This means that it is essential to introduce a deionizing step between saccharification and isomerization so as to remove the calcium ions added to stabilize α-amylase or present in the starch feed. Novo have introduced an α-amylase from *Bacillus licheniformis* which is not dependent on Ca ions (Aschengreen, 1975). It is conceivable that the use of this enzyme in the initial starch-thinning process may allow fructose-enriched glucose syrups to be produced using only a single deionizing step after the isomerization. Such a process has been described by Hollo *et al.* (1975) who passed 18 DE material produced from corn starch using an α-amylase described as 'Tenase' and containing 0·2% $MgCl_2$ and 0·02% $CoCl_2$ at pH 6·5 through three sets of alternate columns of AG and GI immobilized on DEAE-cellulose. The final product had a DE of 96 and contained 42% (on solids) of fructose.

7.5.1 The Need for Cobalt

An objective of research into the production of high-fructose glucose syrups has been to minimize the amount of cobalt ions which must be added to the glucose syrup feedstuff to stabilize and preserve the activity of the GI. This problem does not arise when the GI from *Arthrobacter* species is used (Lee *et al.*, 1972, R. J. Reynolds Tobacco Co., 1973) since the enzyme does not require cobalt to maintain its activity.

Satisfactory results have been achieved when GI has been immobilized, presumably because the cobalt ions are less easily 'leached' from the fixed enzyme. Thus Zittan *et al.* (1975) describe the isomerization of glucose syrup containing only 4×10^{-4} M magnesium ions by passing it through a column of their immobilized GI. Messing and Filbert (1975) state that cobalt ions may be omitted from the glucose syrup feed to GI immobilized on controlled pore alumina once the enzyme has been equilibrated with cobalt. In practice it is probable that the half-life of immobilized GI may be longer if cobalt ions are retained in the feed stuff.

7.6 PSICOSE

D-psicose (D-allulose) is a non-metabolizable ketohexose which is produced in small amounts when glucose and fructose solutions are heated in alkaline conditions (see Section 7.2.1). It is not clear whether the formation of psicose during the production of HFGS is purely chemical or whether it is favoured by the presence of GI. CPC International (1972) demonstrated that when 1 GIU of the GI from *Streptomyces olivochromogenes* was used per gram of solids in 95 DE glucose syrup at 65 °C and pH 7·0 the equilibrium reaction mixture contained 4·4% (w/w solids) of psicose but if 0·77 GIU were used in similar conditions only 1·7% of psicose was produced. Even less psicose was formed when the lower GI level was used at pH 6·25. CPC International (1972) patented a process in which the isomerization was done in an atmosphere of inert gas with the aim of lowering the amount of psicose produced by preserving the life of the (soluble) enzyme and thus allowing the reaction to be conducted at lower pH values. Khaleeluddin *et al.* (1974), however, showed that their fructose-enriched glucose syrups, produced using an unspecified GI, contained no psicose and that no psicose was formed during the refining of that syrup. MKC (1974) recommend that their GI produced from *Streptomyces albus* (Optisweet-P) should not be used above pH 6·5 to prevent the occurrence of non-specific chemical isomerizations.

It seems likely that psicose is formed non-enzymically and that it will not be a problem where immobilized GI is employed in continuous processes where reaction times are short and where the pH of the isomerized syrup is rapidly lowered, e.g. by ion-exchange, after the reaction. The 'Isomerose' produced by Clinton Corn Processing Co. using beds of immobilized GI in a continuous process contains less than 0·4% psicose whereas batch-produced material contains 1·0–1·4% psicose (on solids) (Schnyder, 1974).

7.7 THE TECHNOLOGY OF THE USE OF GLUCOSE ISOMERASE

The processes used by the various manufacturers to produce high-fructose glucose syrups are, not surprisingly, subject to some secrecy. However, Clinton Corn Processing Company's process has been described in some detail (Harden, 1972, 1973; Schnyder, 1974; Mermelstein, 1975) (Figure 7.6). The substrate feed is a syrup containing 93% D-glucose (on a dry substance basis) which has been refined by carbon and ion-exchange, adjusted to the desired pH and supplied with the correct levels of magnesium and cobalt ions. The enzyme, immobilized on DEAE-celluose, is contained in shallow beds (1–5 in. deep) to minimize compaction and pressure drop. Recently the A. E. Staley Manufacturing Co., which uses the Clinton process under licence, has begun to use the Novo immobilized enzyme (Zittan *et al.*, 1975). Two or more beds are used in series in order to minimize the problem of

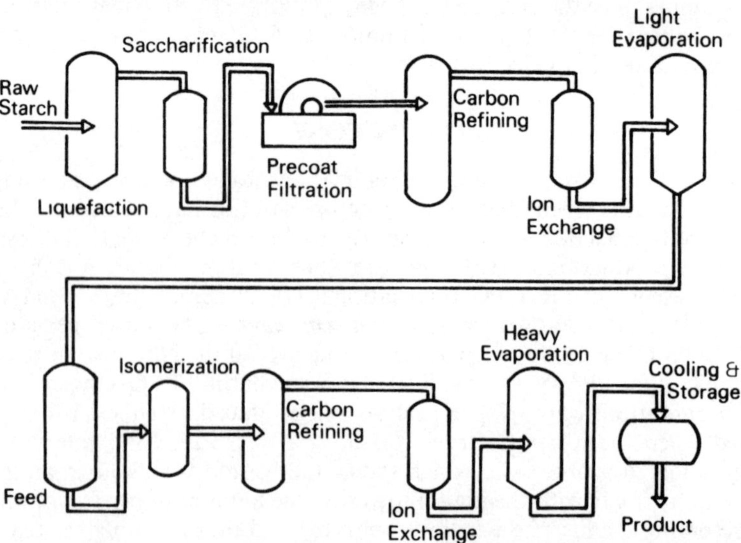

FIG. 7.6 A representative flow chart for the production of high fructose glucose syrup. [*Mermelstein, 1975.*

channelling and to facilitate the replacement of beds of spent enzyme. The isomerized syrup is adjusted to an acidic pH, passed through activated carbon, then through a strong acid cation exchange resin (H^+ form) and finally through a weak base anion exchange resin in the free base form. This sequence removes colour, colour precursors and substantially all the salts. The syrup is concentrated to 71% solids by evaporation and stored. Harden's papers (1972, 1973) describe the computer control of the whole process.

The Anheuser-Busch process has been described in outline (Scallet *et al.*, 1974). This employs the GI from *Actinoplanes missouriensis* batchwise to isomerize 95–98 DE syrup refined by carbon treatment and ion-exchange, adjusted to pH 7·0–7·2 and supplied with Mg and Co ions. The whole *Actinoplanes* cells are used as the enzyme preparation. These are removed by filtration before the isomerized syrup is refined by carbon and ion-exchange treatment and evaporated to 71% solids.

Zittan *et al.* (1975) described the use of the enzyme from *Bacillus coagulans* immobilized for batch use and for continuous use. For batch use, the feed-stuff was adjusted to pH 6·0–7·0 and supplied with $3{\cdot}5 \times 10^{-4}$ M Co^{2+} and 8×10^{-3} M Mg^{2+}. Sufficient enzyme was used to complete the isomerization in 20 hours at 60 °C using a stirred tank reactor. Stirring ceased when isomerization was complete and the enzyme was allowed to settle for 4 hours. Most of the isomerized syrup was then pumped out, leaving the enzyme under a protective layer of syrup to prevent atmospheric oxidation and ready to isomerize another batch of syrup. The activity of the GI decreased after 5–6 uses, retaining 60% of the original activity after 14 cycles of use.

When the more advanced immobilized GI was used in a continuous system the much shorter contact time allowed the use of both a higher temperature (65 °C) and a higher pH (8·5). Also, the need to include cobalt ions in the feed syrup was eliminated and the magnesium ion level could be reduced to one-twentieth of that required in the batch process, thus reducing the cost of purifying the isomerized syrup. It is claimed that decolorization by carbon alone is sufficient to produce an acceptable syrup. Although a high temperature and pH is used, the residence time is too short for detectable amounts of psicose to be formed. Table 7.7 compares the performances of the enzyme used batchwise and continuously.

To summarize, the best process which can be envisaged for the production

Table 7.7 Comparison Between Batch Re-use and Continuous Operation of Novo Immobilized Glucose Isomerase. *(Zittan* et al. *(1975)*.]

	Batch	*Continuous*
Reactor volume	750 m^3	30 m^3
Enzyme consumption	17–20 tons	10 tons
$MgSO_4 \cdot 7H_2O$ consumption	43 tons	2·2 tons
$CoSO_4 \cdot 7H_2O$ consumption	2·2 tons	0
Colour formation $OD_{420\,nm}$	0·05–0·10	0–0·02
Psicose formation	0·1%	0·1%
Product refining	Carbon treatment cation and anion exchange	Carbon treatment

Based on a monthly production of 10,000 tons D.S. 42% fructose.

of HFGS at present would be entirely continuous, employing enzyme-thinned starch saccharified to 95+ DE using immobilized AG and isomerized using a GI immobilized in a form allowing rapid flow rates. The GI would require no cobalt and low levels of magnesium and would be resistant to inhibition by calcium ions, thus allowing the deionization step between saccharification and isomerization to be eliminated.

7.8 THE PRODUCT

The de-ionized, concentrated high-fructose glucose syrup, containing 71% solids, has a bright, haze-free clarity and a 'water-like' colour. It can be handled in apparatus similar to that used for invert sugar or glucose syrups and must be stored at 30–35° to prevent crystallization (Robinson, 1975). Syrups produced from 63 or 75 DE material can be concentrated to 85% solids.

The 'standard' HFGS, containing 42% fructose, is similar in sweetness to sucrose syrup and it has no flavours or odours which might mask those of other ingredients. It can therefore replace sucrose or invert sugar syrups in many food products. There are subtle differences in the sweetness character due to the presence in the HFGS of oligosaccharides and to differences in the texture of the syrups. The high osmotic pressure of the HFGS provides resistance to microbial growth.

The uses to which HFGS has been put are legion and a full description of them would require a separate chapter. This information is given in some detail by Robinson (1975), Gussow (1975), Felton (1975) Godzicki (1975) and many others. Similarly, the fascinating details of the commercial, economic and even social implications of the discovery and use of glucose isomerase are too complex to be dealt with here. An admirable account of the present state of the HFGS industry and of the relationships between the syrup manufacturers, enzyme researchers and enzyme manufacturers is given in Eurofoods (1975).

REFERENCES

Agency of Industrial Science and Technology (1968), British Patent 1,103,394.

Agency of Industrial Science and Technology (1974), British Patent 1,361,846.

Anheuser-Busch Inc. (1975), British Patent 1,399,408.

Aschengreen, N. H. (1975), *Process Biochem.*, **10**(4), 17–19.

Barker, S. A., Somers, P. J. and Hatt, B. W. (to Boehringer Mannheim GmbH) (1975), U.S. Patent 3,875,140.

Baxter Laboratories, Inc. (1972), British Patent 1,274,158.

Baxter Laboratories, Inc. (1975), British Patent 1,401,946.

Bengtson, B. L. and Lamm, W. R. (to Standard Brands Inc.) (1974), British Patent 1,368,511.

Birch, G. G. and Kheiri, M. S. A. (1970), in *Glucose Syrups and Related Carbohydrates*, 31–45. Ed. by G. G. Birch, L. F. Green and C. B. Coulson, Elsevier.

Cotter, W. P., Lloyd, N. E. and Hinman, C. W. (to Standard Brands Inc.) (1971), U.S. Patent 3,623,953.

CPC International Inc. (1972), British Patent 1,295,407.
CPC International Inc. (1975a), British Patent 1,411,763.
CPC International Inc. (1975b), British Patent 1,411,764.
CPC International Inc. (1975c), British Patent 1,411,765.
Danno, G. (1970a), *Agr. Biol. Chem.*, **34**, 1795–1804.
Danno, G. (1970b), *Agr. Biol. Chem.*, **34**, 1805–1814.
Danno, G. (1971), *Agr. Biol. Chem.*, **35**, 997–1006.
Danno, G., Yoshimura, S. and Natake, M. (1967), *Agr. Biol. Chem.*, **31**, 284–292.
Dinelli, D., Morisi, F., Giovence, S. and Pansolli, P. (to Snam Progetti S.p.A.) (1974), British Patent 1,361,963.
Dworschack, R. G., Chen, J. C., Lamm, W. R. and Davis, L. G. (to Standard Brands Inc.) (1972), British Patent 1,284,218.
Dworschack, R. G. and Lamm, W. R. (to Standard Brands Inc.) (1972), British Patent 1,274,640.
Emery, A. N., Novais, J. M. and Barker, S. A. (to Ranks, Hovis and McDougal, Ltd) (1974), British Patent 1,346,631.
Eurofoods (1975), No. 54 (20th November) International Section 1–9.
Felton, T. (1975), *Food Processing Ind.*, **44**(11), 30–32.
Godzicki, M. M. (1975), *Food Eng.*, **47**(10), E.F. 11–17.
Gussow, D. (1975), *Candy Snack Ind.*, **140**(5), 36–42.
Hamilton, B. K., Colton, C. K. and Cooney, C. L. (1974), in *Immobilised Enzymes in Food and Microbial Processes*, 85–131. Ed. by A. C. Olson and C. L. Cooney, Plenum Press.
Harden, J. D. (1972), *Food Eng.*, 44(12), 59–62.
Harden, J. D. (1973), *Food Eng.*, **45**(1), 65–67.
Heady, R. E. and Jacaway, W. A., Jr (to Standard Brands, Inc.) (1974), U.S. Patent 3,847,740.
Hollo, J., Laszlo, E. and Hoschke, A. (1975), *Die Starke*, **27**, 232–235.
Iisuka, H., Ayukawa, Y., Suekane, N. and Kanno, M. (to CPC International Inc.) (1972), British Patent 1,273,003.
Kent, C. A. and Emery, A. N. (1973), *J. Appl. Chem. Biotechnol.*, **23**, 689–703.
Kent, C. A. and Emery, A. N. (1974), *J. Appl. Chem. Biotechnol.*, **24**, 663–676.
Khaleeluddin, K., Sutthoff, R. F. and Nelson, W. J. (to Standard Brands Inc.) (1974), British Patent 1,359,236.
Kolarik, M. J., Chen, B. J., Emery, A. H., Jr and Lim, H. C. (1974), in *Immobilised Enzymes in Food and Microbial Processes*, 71–83. Ed. by A. C. Olson and C. L. Cooney, Plenum Press.
Lamm, W. R., Davis, L. G. and Dworschack, R. G. (to Standard Brands Inc.) (1974) U.S. Patent 3,821,082.
Lee, C. K., Hayes, L. E. and Long, M. E. (to R. J. Reynolds Tobacco Co.) (1972) U.S. Patent 3,645,848.
Lloyd, N. E., Lewis, L. T., Logan, R. M. and Patel, D. N. (to Standard Brands Inc.) (1974), U.S. Patent 3,817,832.
MacAllister, R. V., Lloyd, N. E., Dworschack, R. G. and Nelson, W. J. (to Standard Brands Inc.) (1972), British Patent 1,267,119.
Marshall, R. O. (to Corn Products Co.) (1960), U.S. Patent 2,950,228.
Marshall, R. O. and Kooi, E. R. (1957), *Science*, **125**, 648–649.
Mermelstein, N. H. (1975), *Food Tech.*, **29**(6), 20–22, 24, 26.

Messing, R. A. (to Corning Glass Works) (1974), U.S. Patent 3,850,751.
Messing, R. A. (to Corning Glass Works) (1975), U.S. Patent 3,868,304.
Messing, R. A. and Filbert, A. M. (1975), *J. Agric. Food Chem.*, **23**, 920–923.
Miles Laboratories Inc. (1972), British Patent 1,280,396.
Miles Laboratories Inc. (1974a), British Patent 1,376,787.
Miles Laboratories Inc. (1974b), British Patent 1,376,983.
MKC (Miles Kali-Chemie GmbH & Co. KG) (1974), *Optisweet-P Product Information.*
Monsanto Co. (1974), British Patent 1,356,283.
Moskowitz, G. J. (to Baxter Laboratories Inc.) (1974), U.S. Patent 3,843,442.
Natake, M. (1966), *Agr. Biol. Chem.*, **30**, 887–895.
Natake, M. (1968), *Agr. Biol. Chem.*, **32**, 303–312.
Natake, M. and Yoshimura, S. (1963), *Agr. Biol. Chem.*, **27**, 342–348.
Natake, M. and Yoshimura, S. (1964a), *Agr. Biol. Chem.*, **28**, 505–509.
Natake, M. and Yoshimura, S. (1964b), *Agr. Biol. Chem.*, **28**, 510–516.
Nielsen, T. K. and Markussen, E. K. (1974a), British Patent 1,361,387.
Nielsen, T. K. and Markussen, E. K. (1974b), British Patent 1,362,365.
Research Corporation (1975), British Patent 1,385,585.
R. J. Reynolds Tobacco Co. (1973), British Patent 1,328,970.
R. J. Reynolds Tobacco Co. (1974), British Patent 1,368,650.
Robinson, J. W. (1975), *Food Eng.*, **47**(5), 57–61.
Scallet, B. L. and Ehrenthal, I. (to Anheuser Busch Inc.) (1967), U.S. Patent 3,305,395.
Scallet, B. L. and Ehrenthal, I. (to Anheuser-Busch Inc.) (1968), U.S. Patent 3,383,245.
Scallet, B. L., Katz, E. and Ehrenthal, I. (to Anheuser-Busch Inc.) (1972), U.S. Patent 3,690,948.
Scallet, B. L., Shieh, K., Ehrenthal, I. and Slapshak, L. (1974), *Die Starke*, **26**, 405–408.
Schnyder, B. J. (1974), *Die Starke*, **26**, 409–412.
Shieh, K. K., Donnelly, B. J. and Lee, H. A. (to Anheuser-Busch Inc.) (1974a), U.S. Patent 3,813,320.
Shieh, K. K., Lee, H. A. and Donnelly, B. J. (to Anheuser-Busch Inc.) (1974b), U.S. Patent 3,834,988.
Speck, J. C., Jr (1958), *Adv. Carbohyd. Chem.*, **13**, 63–103.
Standard Brands Inc. (1972), British Patent 1,267,496.
Standard Brands Inc. (1975), British Patent 1,400,829.
Takasaki, Y. (1966), *Agr. Biol. Chem.*, **30**, 1247–1253.
Takasaki, Y. (1967), *Agr. Biol. Chem.*, **31**, 309–313.
Takasaki, Y., Kosugi, Y. and Kanbayashi, A. (1969a), in *Fermentation Advances*, 561–589. Ed. by D. Perlman, Academic Press.
Takasaki, Y., Kosugi, Y. and Kanbayashi, A. (1969b), *Agr. Biol. Chem.*, **33**, 1527–1534.
Thompson, K. N., Johnson, R. A., Lloyd, N. E. (to Standard Brands Inc.) (1974), British Patent 1,371,489.
Tsumura, N. and Sato, T. (1960), *Bull. Agric. Chem. Soc. Japan*, **24**, 326–327.
Tsumura, N. and Sato, T. (1961), *Agr. Biol. Chem.*, **25**, 616–619.
Tsumura, N. and Sato, T. (1965), *Agr. Biol. Chem.*, **29**, 1123–1128.
Tsumura, N., Hagi, M. and Sato, T. (1967), *Agr. Biol. Chem.*, **31**, 902–907.

Vieth, W. R., Wang, S. S. and Saini, R. (1973), *Biotechnol. Bioeng.*, *15*, 565–569.
Weber, P. (to L. Givaudan and S.A. Cie) (1975), British Patent 1,410,579.
Wiseman, A. (1975), in *Handbook of Enzyme Technology*, 272. Ed. by A. Wiseman, Ellis Horwood, Publisher, Chichester.
Yamanaka, K. (1962a), *Agr. Biol. Chem.*, **26**, 167–174.
Yamanaka, K. (1962b), *Agr. Biol. Chem.*, **26**, 175–179.
Yamanaka, K. (1963a), *Agr. Biol. Chem.*, **27**, 265–270.
Yamanaka, K. (1963*b*), *Agr. Biol. Chem.*, **27**, 271–278.
Yamanaka, K. (1965), Japanese Patent 20,230/65.
Yamanaka, K. (1968), *Biochem. Biophys. Acta*, **151**, 670–680.
Yamanaka, K. and Higashihara, T. (1962), *Agr. Biol. Chem.*, **26**, 162–166.
Yoshimura, S., Danno, G. and Natake, M. (1966), *Agr. Biol. Chem.*, **30**, 1015–1023.
Zittan, L., Poulsen, P. B. and Hemmingsen, S. H. (1975), *Die Starke*, **27**, 236–241.

Chapter **8**

Microbial Cytochromes P-450: Drug Applications

Dr ALAN WISEMAN, Department of Biochemistry,
University of Surrey, Guildford, Surrey

8.1 INTRODUCTION

Cytochrome P-450 is best known as the generic name for a complex mixture of enzymes in liver that are responsible for the metabolism of drugs and other xenobiotics (Parke, 1975). It is involved also, in particular steps in the biosynthesis of cholesterol in liver and of steroid hormones in adrenal gland mitochondria.

The cytochrome P-450 of liver is a broad specificity mixed function terminal oxidase. It uses molecular oxygen and reduced cofactor (reduced nicotinamide adenine dinucleotide phosphate, NADPH), e.g. to hydroxylate or dealkylate foreign chemicals prior to excretion. Electrons are supplied by the reduced cofactor through an electron transport carrier chain that includes the flavoprotein, cytochrome P-450 reductase. The system is membrane-bound and is isolated in the microsomal fraction of liver after homogenization and ultracentrifugation.

The feature of cytochrome P-450 which by definition is diagnostic for the presence, and presumed activity of this enzyme, is the characteristic absorption peak at 450 nanometres (nm), observed as a difference spectrum in the presence of carbon monoxide in the test cuvette only. Sodium dithionite is added to both cuvettes of the recording spectrophotometer so that the haem iron present in the enzyme is fully in its reduced (ferrous) form. Very many proteins and enzymes give a peak of absorption at 420 nm under these conditions—the peak at 450 nm is, however, of unusual mechanistic significance and has attracted much attention whenever observed, i.e. throughout animal (including insects), plant and microbial organisms (although it is clear that many other enzymes carry out hydroxylation reactions also). Indeed, its widespread occurrence in nature suggests a general and fundamental role for this enzyme, which might be associated with the removal of e.g. radiation produced free-radicals from the cell.

8.2 CYTOCHROMES P-450 OF PSEUDOMONADS

Strains of *Pseudomonas putida* that grow on the terpene D-(+) camphor as the sole carbon source, produce relatively large amounts of an intracellular and non-bound (to membranes) form of cytochrome P-450. This enzyme is therefore inducible by the substrate, D-camphor, and allows growth on it through its ability to catalyse the first step in its metabolism, i.e. the hydroxylation of the 5-methylene carbon atom to form the exo-5-alcohol, using molecular oxygen and NADH as the electron donor (Hedegaard and Gunsalus, 1965).

This cytochrome P-450 has been subject to large scale purification and indeed has been crystallized, as a substrate complex with D-camphor, in which form it is stabilized (Yu and Gunsalus, 1970). This enzyme was found to be homogeneous to sedimentation in the ultracentrifuge, and contains one mole of ferriprotoporphyrin IX per mole of protein (molecular weight 46,000). The enzyme is usually referred to as P-450_{cam} and is unusual in being highly specific (for a limited range of camphor derivatives), unlike the cytochromes P-450 of yeasts. It must be recognized, however, that the broad specificity of many of these other cytochromes P-450 may be due to the presence of extreme heterogeneity that is difficult to detect when starting with membrane-bound species. Therefore, despite the ready availability of this form of cytochrome P-450, an application in enzyme technology must seem unlikely, at least in its ability to hydroxylate D-camphor and some of its derivatives—unless it were coupled to some useful reaction generating oxygen or organic peroxides.

Cytochrome P-450 has been subjected to large scale purification and extensive study (e.g. Peterson, 1971; Yu and Gunsalus, 1974; Yu *et al.*, 1974). The molecule was reported to be a single polypeptide chain, with molecular weight 44,000–46,000. Spectroscopic studies have provided evidence for particular mechanisms of action of this enzyme (review; Griffin *et al.*, 1975, also Gunsalus *et al.*, 1974).

A structural resemblance has been reported between P-450_{cam} and the cytochrome P-450 from rabbit liver microsomes (Dus *et al.*, 1975). Highly purified forms of these enzymes show immunological cross reactions by competitive binding and inhibition of catalytic activity. They are also of similar sub-unit molecular weight and amino acid composition. Thus, although they are very different in substrate specificity and solubility, they are similar structurally. For its action in addition to the flavoprotein reductase component of the electron-transport chain, P-450_{cam} requires the iron–sulphur protein, putidaredoxin; while the mammalian enzyme requires a phospholipid (associated with its membrane location).

Further studies on the mechanism of P-450_{cam} in relation to other enzymes of this family are of considerable interest (Gunsalus *et al.*, 1976).

It should be noted that enzymes other than the form recognized spectrally as cytochrome P-450 are present in micro-organisms including Pseudomonads, which are involved in the metabolism of suitable growth substrates in the role of terminal oxidase. For example, a carbon-monoxide insensitive alkane hydroxylase, acting for example on n-octane, has been reported in *P. oleovorans* Peterson *et al.*, 1966, 1967, 1968). *P. aminovorans* has been reported to achieve the N-demethylation of certain secondary amines used as a carbon source. This enzyme exhibits a difference spectra with peak at 420 nm rather than 450 nm (Eady *et al.*, 1971).

In addition, many micro-organisms achieve biotransformations of drugs, often by poorly understood mechanisms, probably not involving cytochrome P-450 as the terminal oxidase. These include aromatic hydroxylations, N- and O-dealkylations and sulphur oxygenations. A thorough review of this field has been published (Smith and Rosazza, 1975). Nevertheless, cytochrome P-450 containing systems are of particular interest in view of their similarities with, and differences from, the mammalian enzymes.

8.3 CYTOCHROMES P-450 OF OTHER BACTERIA

A well established role for cytochrome P-450 is in the oxidation of n-octane by *Corynebacterium spp.* This enzyme was detected spectrophotometrically, in soluble form after disruption of the cells by ultrasonication. It required molecular oxygen and NADH and was inhibited by carbon monoxide, and not by cyanide (Cardini and Jurtshuk, 1968, 1970), as other cytochromes P-450. This contrasts with the *P. oleovorans* terminal hydroxylase system for n-octane oxidation (see above).

An interesting involvement of cytochrome P-450 is in the maintenance of the anaerobic nitrogen-fixing bacteroid of the legume *Rhizobium japonicum*, grown symbiotically in soybean root nodules (Daniel and Appleby, 1972). Respiration might occur possibly through a system, requiring cytochrome P-450 as the terminal electron acceptor, at low oxygen tension.

8.4 CYTOCHROMES P-450 OF FUNGI, OTHER THAN YEASTS

Here *Claviceps purpurea* is known to produce this enzyme for the metabolism of alkaloids (Wilson and Orrenius, 1971). It is of particular interest that the cytochrome P-450 in this organism has been shown to be inducible by phenobarbital (Ambike *et al.*, 1970), comparable with the well-known effect of injected phenobarbital in inducing the biosynthesis of cytochrome P-450 in liver (Parke, 1975). Such an effect has been demonstrated also for brewers' yeast (Wiseman and Lim, 1975a).

Cytochrome P-450-like activity has been reported also in a fungal organism, *Cunninghamella bainieri* (Ferris *et al.*, 1973). Many drugs could be metabolized, including the N-demethylation of aminopyrine, the O-dealkylation of p-nitroanisole and the 4-hydroxylation of aniline. Another cytochrome P-450 capable of p. O-dealkylation reactions has been isolated and purified from *Nocardia spp* (Broadbent and Cartwright, 1974). This enzyme had a molecular weight of 42,000–45,000, as for the *P. putida* enzyme.

8.5 CYTOCHROMES P-450 OF YEASTS, OTHER THAN *SACCHAROMYCES spp*

A species of *Torulopsis* when grown on a medium containing glucose, has been reported to hydroxylate added fatty acids at the terminal or penultimate carbon atoms, depending on chain length and degree of unsaturation (Tulloch *et al.*, 1962). Hydrocarbons can also be hydroxylated, in some cases with chain-shortening. Stearic and oleic acids are hydroxylated mainly at the penultimate carbon atom, giving 17-L-hydroxy acids as the product. Further studies with this yeast system suggested the involvement of cytochrome P-450; e.g. it required O_2 and NADPH and was inhibited by carbon monoxide, but not cyanide (Heinz *et al.*, 1970).

A seemingly similar system has been extensively studied from a strain of *Candida tropicalis*, grown on tetradecane (substrate induction). The involvement of cytochrome P-450 in this system was reported about the same time by two groups of workers (by Gallo *et al.*, 1971 and by Lebeault *et al.*, 1971). The enzyme may be solubilized after extensive disruption using a French pressure cell, although it is likely to be located in the microsomal fraction of the cell (Gallo *et al.*, 1971). The ω-hydroxylation of fatty acids (especially lauric acid) and the analogous conversion of n-alkanes to primary alcohols (hydroxylation at the methyl carbon atoms in both cases), were demonstrated in an oxygen-dependent reaction that requires NADPH. (NADH will increase the activity if added along with saturating levels of NADPH.) The complete enzyme system contained, in addition to the cytochrome P-450, the NADPH-cytochrome P-450 reductase (this can be assayed as a cytochrome c reductase) plus a heat-stable lipid fraction (Duppel *et al.*, 1973). It is of great interest that these two components could be replaced by the corresponding fractions obtained from rat liver—which also catalyses fatty and ω-hydroxylation, as does rat kidney cortex (Ellin and Orrenius, 1973). It is of interest in comparison with the Torulopsis yeast (see above) that stearic acid was reported to be ω-hydroxylated by the Candida yeast system. A list of eleven compounds tested was given (Lebeault *et al.*, 1971).

The hydroxylating ability of these cytochrome P-450 containing systems from these yeasts may be of future industrial value, depending on their specificity. In addition, the N-demethylation of certain drugs (aminopyrine, hexobarbital, benzphetamine and ethylmorphine) was demonstrated (Lebeault *et al.*, 1971).

8.6 CYTOCHROMES P-450 OF *SACCHAROMYCES spp* OF YEASTS

This enzyme was detected first, in baker's yeast, by Lindenmayer and Smith (1964). They found this enzyme in yeast grown anaerobically in 10% glucose medium supplemented (for faster growth) with ergosterol and oleic acid. The enzyme was present also in yeast grown aerobically in 4% glucose medium (the yeast was harvested at about the end of the exponential phase of growth).

This work was continued in Japan. The low yield of cytochrome P-450 under aerobic conditions (using 1% glucose medium) was emphasized (Ishidate *et al.*, 1969a). More detailed studies by these workers (Ishidate *et al.*, 1969b) revealed that the cellular content of cytochrome P-450 in yeast grown anaerobically decreased sharply over four hours on exposure to aerobic conditions. The enzyme is presumably degraded or converted to some other cytochrome, perhaps during the biogenesis of functional mitochondria. After anaerobic growth, yeast 'mitochondria' lack the cytochrome $a+a_3$ needed for aerobic respiration, but rapidly form this cytochrome upon aeration. These authors noted the parallel loss of cytochrome P-450 with the start of functioning of mitochondria, as indicated by the high rate of respiration (μmoles O_2 absorbed/minute) and the appearance of cytochrome a.

High concentrations of glucose or of chloramphenicol prevented the development of mitochondrial respiration and also prevented the loss of the cytochrome P-450. It was possible to demonstrate an uptake of oxygen in the presence of antimycin A, which was therefore non-mitochondrial. This antimycin-insensitive respiration was proportional to the cytochrome P-450 content of the yeast under a variety of conditions in wild type yeast and respiratory-deficient mutants (which accumulate this enzyme even in low glucose aerobically grown yeast). These respiratory-deficient mutants accumulate much larger amounts (threefold) of cytochrome P-450 than the wild type under (semi) anaerobic conditions. This level also drops sharply on aeration even though no cytochrome a formation can be detected here.

On this evidence, it was thought unlikely that cytochrome P-450 is converted to cytochrome a on aeration, and concomitant mitochondriogenesis. Ishidate concluded it was more likely that this enzyme is involved in a system that catalyses an oxygen-dependent transformation of lipids needed for formation of mitochondrial membrane; the cytochrome P-450 disappeared, presumably during this process. The subcellular location of the enzyme was not established in this work. In subsequent work (Kawaguchi *et al.*, 1973), most of the respiratory activity of semi-anaerobically grown yeast was found in the soluble and large particulate fraction (25,000 × g for 20 min) rather than in the small particulate fraction (105,000 × g for 60 min). The yeast was disrupted by the use of snail gut enzyme followed by exposure to hypotonic medium (see e.g. Williams and Wiseman, 1973). Both particulate systems showed similar properties, for example in the use of NADPH (and to some extent NADH) in an antimycin-insensitive respiration, sensitive to inhibition by carbon monoxide. Both of these particulate systems were considered to be analogous to the microsomal fraction of liver. The supernatant system used NADPH (and to some extent NADH) in an antimycin-insensitive respiration that could not be inhibited by carbon monoxide. This was attributed by the authors to an electron transfer system distinguishable from that involving cytochrome P-450 (little cytochrome P-450 was detected spectrophotometrically in this supernatant fraction).

Previous workers have claimed that both NADPH and molecular oxygen are required for the synthesis of ergosterol and for the desaturation of fatty acids in yeast (Bloomfield and Bloch, 1960; Klein, 1960), but not in similar cases with liver (Yamamoto and Bloch, 1970). It is for this reason that ergosterol and Tween-80 (an oleic acid derivative) are added to assist the strict anaerobic growth of yeast (Lindmeyer and Smith, 1964; Ferdouse *et al.*, 1972). Nevertheless in yeast grown semi-anaerobically, without the addition of ergosterol or Tween-80 and yet under an atmosphere of carbon monoxide sufficient to convert 98% of the cytochrome P-450 into its CO complex, there was no inhibition of growth rate (semi-anaerobic) or lowering of the ergosterol or phospholipid phosphorus levels in the yeast (Kawaguchi *et al.*, 1973). The authors concluded that cytochrome P-450 is not involved in ergosterol synthesis, nor in the desaturation required for the formation of the unsaturated fatty acid moiety in yeast phospholipids, unless saturated fatty acids could replace these, under such conditions. This conclusion is contrary to studies on a cell-free system from *Saccharomyces cerevisiae* that can convert lanosterol to zymosterol (Alexander *et al.*, 1974). This system, in the pathway for ergosterol biosynthesis, involves a demethylation reaction, and it was inhibited 57% by carbon monoxide. The authors suggested the involvement therefore of cytochrome P-450 in this pathway. It is possible that carbon monoxide is incapable of penetrating to the site of cytochrome P-450 binding in membranes, in the whole yeast. Kawaguchi *et al.* (1973) also noted the interesting possibility that the cytochrome P-450 of these yeasts may be involved in ω-hydroxylation of fatty acids as reported for *Candida tropicalis* (Lebeault *et al.*, 1971; Gallo *et al.*, 1971) and as reported for *Torulopsis magnoliae* grown on glucose supplemented with long-chain fatty acids (Heinz *et al.*, 1970; also Tulloch *et al.*, 1962). In the latter case, a 12,000 × g supernatant fraction of the yeast was able to hydroxylate many fatty acids and hydrocarbons, including the conversion of oleic acid to 17-L-hydroxyoleic acid.

Kawaguchi *et al.* (1973) however, considered such ω-hydroxylations to be unlikely in their yeast because of the presence of the cytochrome P-450 in semi-anaerobically-grown yeast and its disappearance upon aerobic adaptation. An alternative view would be that the hydroxylation of oleic acid, although unlikely in *Saccharomyces spp*, might be required under such conditions of mitochondrial repression! Seemingly in support of this view we find that 3% Tween-80 (but not ergosterol) greatly increases the level of cytochrome P-450 in brewer's yeast at late phases of growth (Wiseman *et al.*, 1976); see Figure 8.1. This would be a case of substrate induction, which would suggest that cytochrome P-450 catalyses the hydroxylation of oleic acid prior to incorporation into membranes possibly of the endoplasmic

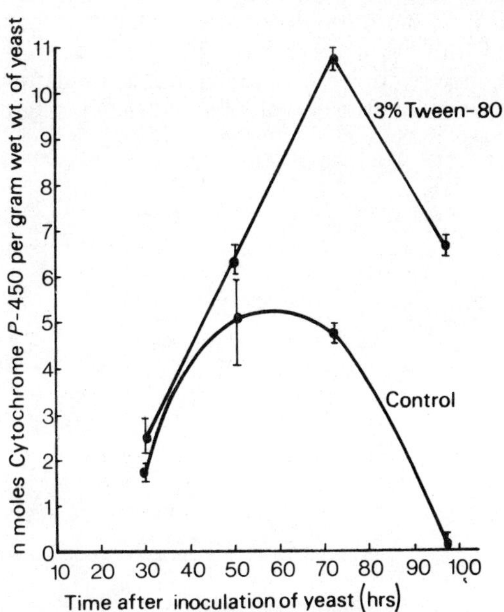

FIG. 8.1 Effect of 3% tween 80 on accumulation of cytochrome P-450 in yeast in 20% glucose medium.

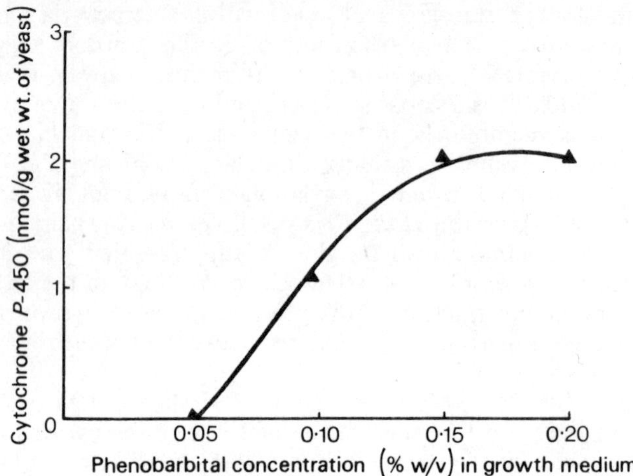

FIG. 8.2 Effect of phenobarbital concentration on accumulation of cytochrome P-450 in yeast in 0·5% glucose medium in 24 hr.

reticulum or, more likely, of the mitochondrion (required to be completed on removal of mitochondrial repression) (see Wiseman, 1975a). Cytochrome P-450 is induced by phenobarbital in our yeast (Wiseman and Lim, 1975) (Figures 8.2, 8.3) but this occurs in 0·5% glucose medium where none of this enzyme is normally formed. No additional enzyme is produced in 20% glucose medium in the presence of phenobarbital. An alternative explanation therefore for the Tween-80 effect might be in the stabilization of the cytochrome P-450 against the normal rapid loss that occurs in late phases of growth associated with the fall of residual glucose levels to below 8%.

We have studied the relationship between the accumulation of cytochrome P-450 in brewer's yeast and the growth phase of the yeast, grown aerobically

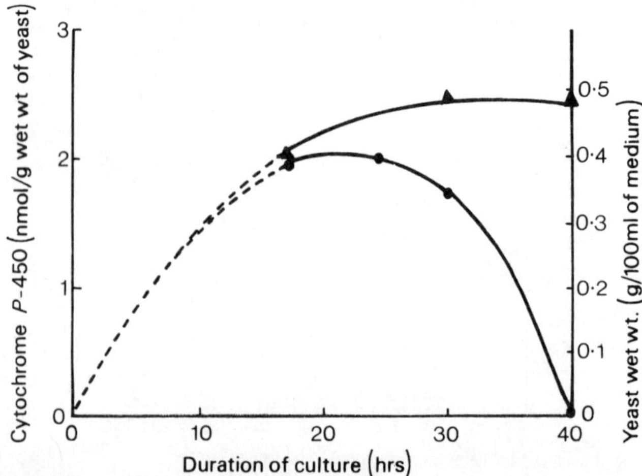

FIG. 8.3 Time-course of cytochrome P-450 accumulation and loss in yeast in the presence of 0·2% phenobarbital in 0·5% glucose medium (●) and the yeast growth curve in 0·5% glucose plus 0·2% phenobarbital (▲).

in glucose-containing media in the range 0·1% to 20% glucose (Wiseman *et al.*, 1975a; also Wiseman *et al.*, 1976). No enzyme is produced in 0·1% glucose medium. Early accumulation of the enzyme was observed in low-glucose media (1%, 2%) and late accumulation was observed in high-glucose media (5% upwards); with intermediary behaviour in 3% or 4% glucose media (Figure 8.4). There is an inverse correlation of cytochrome P-450 level

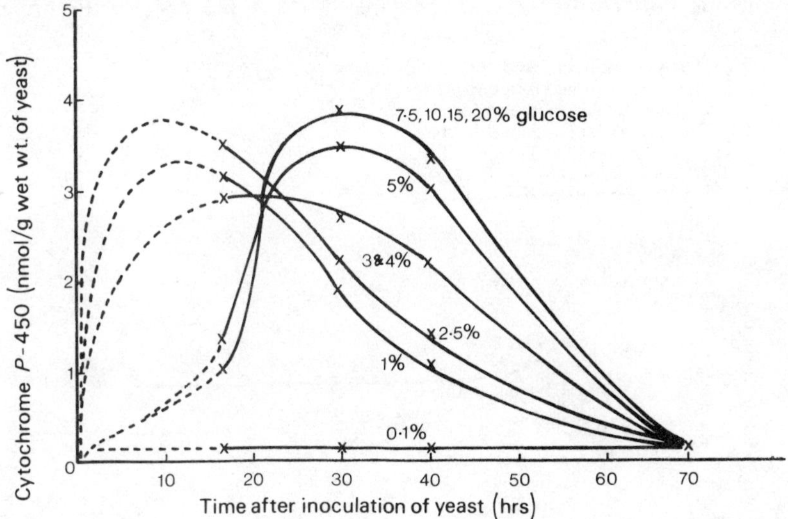

FIG. 8.4 Changes in concentration of cytochrome P-450 in yeast during growth in 0·1% to 20% glucose medium (as indicated).

with that of cytochrome $a + a_3$, and therefore with the level of cyclic AMP in the yeast under a variety of conditions (Wiseman, A. and Lim, T.-K., unpublished work). Other such studies using protoplasts, made from the yeast by the use of snail gut enzyme, have shown that added cyclic AMP enters the protoplast and prevents the accumulation of cytochrome P-450 (and increases its rate of disappearance on transfer to low glucose medium), while cyclic GMP can cause its accumulation in 0·5% glucose medium, where it is otherwise not formed (Wiseman, A. and Lim, T.-K., unpublished work). Our evidence suggests that the transcription of the gene responsible for cytochrome P-450 synthesis in yeast is controlled in the opposite fashion to the classic *LAC* operon of *E. coli* or that for the synthesis of, e.g. α-glucosidase in yeast (see review, Wiseman, A., 1975b).

The drug specificity of our cytochrome P-450 from brewer's yeast has not been fully elucidated. Nevertheless, it is clear that a specific 4′-hydroxylation of biphenyl is catalysed by the microsomal fraction (where most of the enzyme is found), obtained after disruption of the yeast by Vibro Mill or by snail gut enzyme (Wiseman *et al.*, 1975b; also Wiseman *et al.*, 1975c). In addition, we have noted the demethylation of N-ethylmorphine and of aminopyrine, as was reported for the cytochrome P-450 from *Candida tropicalis* (Lebeault *et al.*, 1971). Studies on our biphenyl 4′-hydroxylation system have shown its dependence on NADPH or almost equally on NADH. No synergistic effect was found, contrary to some findings on a soluble enzyme from *Candida tropicalis* (Duppel *et al.*, 1973), but in agreement with the studies on a microsomal enzyme from this yeast (Gallo *et al.*, 1971).

The thermal (and working) stability of cytochromes P-450 are of great interest in relation to the possibility of application (Wiseman *et al.*, 1975b). We have compared the oxidized (Fe^{3+}) and reduced (Fe^{2+}) forms of liver cytochrome P-450 with the yeast enzyme at 46 °C (pH 7·4); see Figure 8.5

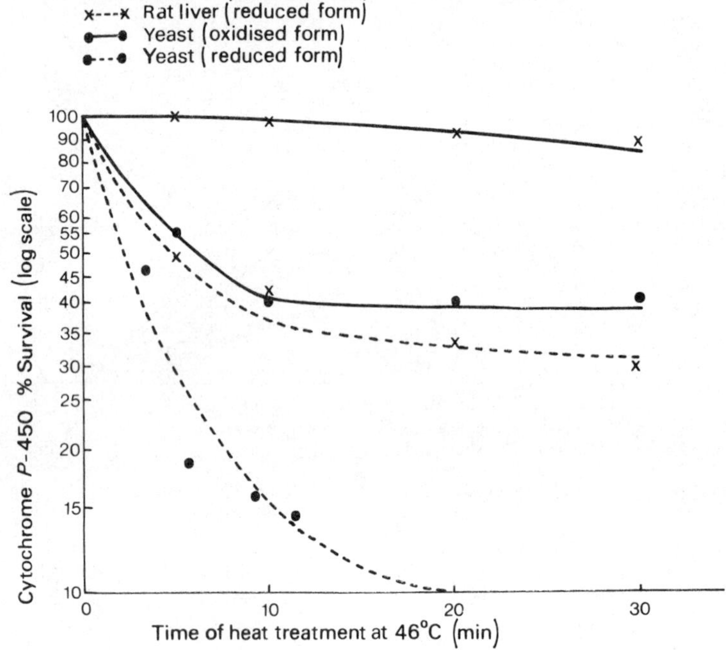

FIG. 8.5 Thermal destruction of cytochrome P-450 at 46 °C in microsomal fraction from rat liver or yeast.

(spectral P-450 loss, agrees with loss of biphenyl hydroxylase activity). The oxidized form of the liver enzyme was almost completely stable over 30 minutes under these conditions, and the reduced form was considerably less stable, much like the oxidized form of the yeast enzyme. It would seem from the biphasic logarithmic plot obtained that at least two species of enzyme are present, one of which is stable. The reduced form of the yeast enzyme was extremely unstable. It is clear that the operational cycle of the enzyme, alternating between oxidized and reduced form in the conversion of drugs, would require additional stabilization by immobilization or cross-linking, especially in the case of the yeast enzyme. The latter can be readily obtained in large amounts, but is less stable than the liver enzyme and may have a much more limited specificity; a disadvantage in some applications that could be envisaged. One hopeful observation is that substrates are capable of protecting the enzyme under some conditions. We have reported (Wiseman and Gondal, 1975) the protection of the enzyme against a detergent (Tween-80)—assisted thermal denaturation by both Type 1 and Type 2 substrate examples (Type 1, N-ethylmorphine: Type 2, aniline). Binding spectra for substrates fall mainly into these two spectral categories when oxidized-oxidized plus substrate, difference spectra are recorded (see review: Parke, 1975). Other protective agents (e.g. in storage) for cytochrome P-450 include glycerol and thiol compounds such as cysteine and dithiothreitol.

Attempts to isolate the soluble form of yeast cytochrome P-450 have been reported (Yoshida and Kumoaka, 1975). Microsomal fraction was treated with a proteolytic enzyme to remove unwanted proteins, and then extracted with sodium cholate, followed by precipitation with ammonium sulphate. The enzyme was purified (overall 30–40-fold) using Sephadex G-25 followed by DEAE-cellulose column chromatography. The spectral characteristics were reported to be identical to the cytochrome P-450 of hepatic microsomes, especially that from polycyclic hydrocarbon-induced animals (i.e. P-448, see Parke, 1975). The enzyme achieved the demethylation of aminopyrine, but unlike our findings it was thought possibly able to hydroxylate aniline. Most of the enzyme was found to be in the high-spin spectral state (probably in equilibrium with the low-spin state, Yoshida and Kumoaka, 1972). The Type 1 spectral change shown by some substrates (see above) was not found (in agreement with our findings with biphenyl, a Type 1 substrate), as this change is due to the transition from the low-spin to high-spin state. Other studies by these authors (Yoshida *et al.*, 1974a, b) showed the presence in the yeast system of the same electron-transferring components as found in liver, i.e. in addition to cytochrome P-450, the cytochrome b_5, the NADH-cytochrome b_5 reductase, and the NADPH-cytochrome c reductase (which donates electrons to cytochrome P-450).

The presence of cytochrome P-450 has been mentioned by other workers, e.g. in *Saccharomyces carlsbergenesis* (Cartledge *et al.*, 1972). It has been reported also in the fission yeast, *Schizosaccharomyces pombe*, especially noted in the last one-quarter of the division cycle. A possible physiological role in respiration was discussed by these authors (Poole *et al.*, 1974).

8.7 APPLICATIONS OF CYTOCHROMES P-450

A large number of drug biotransformations by micro-organisms have been reported (see comprehensive review, Smith and Rosazza, 1975; also Gunsalus *et al.*, 1975). Only a few of these micro-organisms have been shown to contain, or use, cytochrome P-450; although many of the mono-oxygenase systems are remarkably similar in their specificities and mechanisms. It may prove possible in some cases to predict and identify the metabolites likely to be produced by hepatic metabolism of drugs and other foreign compounds, following the isolation of gram quantities of metabolites using microbial cytochromes P-450. These could be subjected to full structural identification and the assessment of toxicological (including carcinogenic) activity.

An alternative approach would be to stabilize the cytochrome P-450 system from liver. An immobilized form, apparently, of the complete system, has been prepared for this purpose (Sofer *et al.*, 1975). Here, a pig liver microsomal fraction was used to prepare a soluble form of cytochrome P-450. This was subsequently bound to zirconia-clad 1350 Å pore diameter glass beads using glutaraldehyde as a carbonyl intermediate (the enzyme was attached by a Schiff-base coupling procedure). An operating range of 25–28 °C was found to be optimal, at pH 7·6. This immobilized enzyme catalyses the oxidation of a variety of amines and hydrazines, but was principally used by these workers to demonstrate the preparation of milligram quantities of the N-oxide metabolites of prochlorperazine, guanethidine and ethylmorphine (all using a simple slurry reactor technique). This immobilized cytochrome P-450 is thermolabile in use—even at 35 °C the half-life was about 10 hours only—but was reasonably stable on storage in buffer at 8 °C. Some microbial forms of the enzyme may prove useful in such an application.

Other applications that can be envisaged for cytochromes P-450 may not be easy or even feasible. These include the possible medical applications either by injection (in encapsulated or liposome lipid globule form to avoid immunological responses), or by use in extracorporeal shunts for liver replacement or drug abuse therapy.

Cytochrome P-450 is used in the well-known Ames test (Ames, 1976). The enzyme (rat liver or human liver forms) is used for pretreatment of drugs to form any toxic metabolites as *in vivo*, when testing for mutagenicity (and therefore possible carcinogenicity) using histidine-lacking mutants of *Salmonella typhimurium*. Mutagens can cause the formation of revertent forms of the bacterium, able to grow on the medium lacking in histidine.

Cytochrome P-450 may be envisaged also as an environmental cleaner in suitable flow-through systems, using aqueous wastes. Many desirable transformations of chemicals including pesticides are possible, although other possibly harmful compounds may be potentiated in this way.

REFERENCES

Alexander, K. T. W., Mitropoulos, K. A. and Gibbons, G. F. (1974), *Biochem. Biophys. Res. Commun.*, **60**, 460–467.

Ambike, S. H., Baxter, R. H. and Zahid, N. D. (1970), *Phytochemistry*, **9**, 1953–1958.

Ames, B. N. (1976), *Science*, **191**, 241–244.

Bloomfield, D. K. and Bloch, K. (1960), *J. Biol. Chem.*, **235**, 337–345.

Broadbent, D. A. and Cartwright, N. J. (1974), *Microbios*, **9**, 119–130.

Cardini, G. and Jurtshuk, P. (1968), *J. Biol. Chem.*, **243**, 6071–6072.

Cardini, G. and Jurtshuk, P. (1970), *J. Biol. Chem.*, **245**, 2789–2796.

Cartledge, T. G., Lloyd, D., Erecinska, M. and Chance, B. (1972), *Biochem. J.*, **130**, 739–747.

Daniel, R. M. and Appleby, C. A. (1972), *Biochim. Biophys. Acta*, **275**, 347–354.

Duppel, W., Lebeault, J.-M. and Coon, M. J. (1973), *Eur. J. Biochem.*, **36**, 583–592.

Dus, K., Litchfield, W. J., Miguel, A. G., Van der Hoeven, T. A., Haugen, D. A., Dean, W. L. and Coon, M. J. (1974), *Biochim. Biophys. Res. Communs.*, **60**, 15–21.

Eady, R. R., Jarman, T. R. and Large, P. J. (1971), *Biochem. J.*, **125**, 449–459.
Elfin, A. and Orrenius, S. (1973), *Arch. Biochem. Biophys.*, **158**, 597–604.
Ferdouse, M., Rickard, P. A. D., Moss, F. J. and Blanch, H. W. (1972), *Biotechnol. Bioeng.*, **14**, 1007–1026.
Ferris, J. P., Fasco, M. J., Stylianopoulou, F. L., Jerina, D. M., Daly, J. W. and Jeffrey, A. M. (1973), *Arch. Biochem. Biophys.*, **756**, 97–103.
Gallo, M., Bertrand, J. C. and Azoulay, E. (1971), *FEBS Lett.*, **19**, 45–49.
Griffin, B. W., Peterson, J. A., Werringloer, J. and Estabrook, R. W. (1975), *Annal. N.Y. Acad. Sci.*, **244**, 107–131.
Gunsalus, I. C., Meeks, J. R., Lipscomb, J. D., Debrunner, P. G. and Munck, E. (1974), in *Molecular Mechanisms of Oxygen Activation.* Ed. by O. Hayaishi, Academic Press.
Gunsalus, I. C., Pederson, T. C. and Sligar, S. G. (1975), *Ann. Rev. Biochem.*, **44**, 377–407.
Gunsalus, I. C., Sligar, S. G. and Debrunner, P. G. (1976), *Biochem. Soc. Trans.*, 3, 821–825.
Hedegaard, J. and Gunsalus, I. C. (1965), *J. Biol. Chem.*, **240**, 4038–4043.
Heinz, E., Tullock, A. P. and Spencer, J. F. T. (1970), *Biochim. Biophys. Acta*, **202**, 49–55.
Ishidate, K., Kawaguchi, K., Tagawa, K. and Hagihara, B. (1969), *J. Biochem.* (Tokyo), **65**, 375–383.
Ishidate, K., Kawaguchi, K. and Tagawa, K. (1969), *J. Biochem.* (Tokyo), 3, 385–392.
Kawaguchi, K., Ishidate, K. and Tagawa, K. (1973), *J. Biochem.* (Tokyo), **74**, 817–826.
Klein, E. P. (1960), *J. Bacteriol.*, **80**, 665–672.
Lebeault, J.-M., Lode, E. T. and Coon, M. J. (1971), *Biochim. Biophys. Res. Communs.*, **42**., 413–419.
Lindenmayer, A. and Smith, L. (1964), *Biochim. Biophys. Acta*, **93**, 445–461.
Parke, D. V. (1975), in *Enzyme Induction.* Ed. by D. V. Parke, Plenum Press, 207–271.
Peterson, J. A. (1971), *Arch. Biochem. Biophys.*, **144**, 678–693.
Peterson, J. A. and Coon, M. J. (1968), *J. Biol. Chem.*, **243**, 329–334.
Peterson, J. A., Basu, D. and Coon, M. J. (1966), *J. Biol. Chem.*, **241**, 5162–5164.
Peterson, J. A., Kusunose, M., Kusunose, E. and Coon, M. J. (1967), *J. Biol. Chem.*, **245**, 4334–4340.
Poole, R. K., Lloyd, D. and Chance, B. (1974), *Biochem. J.*, **138**, 201–210.
Smith, R. V. and Rosazza, J. P. (1975), *Biotech. Bioeng.*, **17**, 785–814.
Sofer, S. S., Ziegler, D. M. and Popovich, R. P. (1975), *Biotech. Bioeng.*, **17**, 107–117.
Tulloch, A. P., Spencer, J. F. T. and Gorin, P. A. J. (1962), *Canad. J. Chem.*, **40**, 1326–1338.
Tyson, C. A., Lipscomb, J. D. and Gunsalus, I. C. (1972), *J. Biol. Chem.*, **247**, 5777–5784.
Williams, N. J. and Wiseman, A. (1973), *Biochem. Soc. Trans*, **1**, 1301–1303.
Wilson, B. and Orrenius, S. (1971), *Chem. Biol. Ints.*, **3**, 313–315.
Wiseman, A. (1975a), *J. Appl. Chem. Biotechnol.*, in the press.

Wiseman, A. (1975b), in *Enzyme Induction.* Ed. by D. V. Parke, Plenum Press, 1–26.

Wiseman, A. and Lim, T.-K. (1975), *Biochem. Soc. Trans.*, 3, 974–977.

Wiseman, A. and Gondal, J. A. (1975), *Biochem. Soc. Trans.*, 3, 977–980.

Wiseman, A., McCloud, C. and Lim, T.-K. (1976), *Biochem. Soc. Trans.*, in the press.

Wiseman, A., Lim, T.-K. and McCloud, C. (1975a), *Biochem. Soc. Trans.*, 3, 276–278.

Wiseman, A., Gondal, J. A. and Sims, P. (1975b), *Biochem. Soc. Trans*, 3, 278–281.

Wiseman, A., Jay, F. and Gondal, J. A. (1975c), *J. Sci. Food and Agricult.*, **26**, 539–540.

Yamamoto, S. and Bloch, K. (1970), *J. Biol. Chem.*, **245**, 1670–1674.

Yoshida, Y. and Kumoaka, H. (1972), *J. Biochem.* (Tokyo), **71**, 915–918.

Yoshida, Y. and Kumoaka, H. (1975), *J. Biochem.* (Tokyo), **78**, 785–794.

Yoshida, Y., Kumoaka, H. and Sato, R. (1974a, b), *J. Biochem.* (Tokyo), **75**, 1201–1210, also 1211–1219.

Yu, C. A. and Gunsalus, I. C. (1970), *Biochem. Biophys. Res. Communs.*, **40**, 1423–1428.

Yu, C. A. and Gunsalus, I. C. (1974), *J. Biol. Chem.*, **249**, 102–106 and 107–110.

Yu, C. A., Gunsalus, I. C., Kataguri, M., Suhara, K. and Takemori, S. (1974), *J. Biol. Chem.*, **249**, 91–101.

Index

Index